Neurosurgical Aspects of Epilepsy

Proceedings of the Fourth Advanced Seminar in Neurosurgical
Research of the European Association of Neurosurgical Societies
Bresseo di Teolo, Padova, May 17–18, 1989

Edited by

J. D. Pickard, G. Maira, C. E. Polkey, T. Trojanowski

Acta Neurochirurgica
Supplementum 50

Springer-Verlag Wien New York

Professor John D. Pickard
Wessex Neurological Centre, Southampton General Hospital, Southampton, U.K.

Professor Giulio Maira
Istituto di Neurochirurgia, Università Cattolica, Roma, Italy

Mr. Charles E. Polkey
Neurosurgical Unit, Maudsley Hospital, London, U.K.

Professor Tomasz Trojanowski
Department of Neurosurgery, University Medical School, Lublin, Poland

With 19 Figures

ISSN 0065-1419
ISBN-13: 978-3-7091-9106-4   e-ISBN-13: 978-3-7091-9104-0
DOI: 10.1007/978-3-7091-9104-0

Preface

The 4th Advanced Seminar in Neurosurgical Research was held on May 17th–18th, 1989 in the Abbey of Praglia, a XIIth century Benedictine Monastery, near Padua, Italy, and was devoted to "Neurosurgical Aspects of Epilepsy". The general aim of these Advanced Seminars is to bring together European Neurosurgeons interested or involved in research, either clinically, experimentally or both, in a given field in order to achieve in-depth informal discussions not possible in the more conventional large congress. In particular, these Advanced Seminars seek to provide high level teaching by experienced basic scientists, to provide "state of the art" assessment of the subject and to highlight areas of controversy that might be suitable for future research. A special effort is made to identify younger Neurosurgeons, partly through the auspices of the European Directory of Neurosurgical Research, who have a particular interest in the subject under discussion, not all of whom will have immediate access to the most advanced, modern technology.

Surgical treatment of epilepsy is an expanding area of endeavour and an expertise that remains underutilized in many countries. The programme for this Seminar was designed to paint the broad background, moving from basic membrane electrophysiology through to cost benefit analysis and rehabilitation. The perspectives of neuropathology, neurology, neuroradiology, nuclear medicine, neuropsychology, neurophysiology and the drug industry are all included. A particular effort was made to provide a broad European view that has achieved a singular poignancy in the light of dramatic recent political events. The Seminar did not seek to cover the same ground so expertly covered by Engel (Surgical Treatment of the Epilepsies – Editor J. Engel, Raven Press, New York, 1987).

We would like to pay particular tribute to Dr. A. Molendini and his staff (Dr. L. Linzi and Miss Mannarino) of the Scientific Public Relations Department of FIDIA Research Laboratories who sponsored and organized this most memorable Seminar. The Benedictine atmosphere of Praglia Abbey will stay with those fortunate enough to participate in this meeting for many years. The Research Committee hopes that the manuscripts of the Seminar will stimulate further research into intractable epilepsy for the benefit of all our patients: the most important function of the Research Committee is to stimulate, facilitate and publicize high quality research by young Neurosurgeons in Europe.

J. D. Pickard
G. Maira
C. E. Polkey
T. Trojanowski

on behalf of the Research Committee of the European Association of Neurosurgical Societies.

October 1990

Contents

Listed in Current Contents

Acta Neurochirurgica, Suppl. 50, 1–5 (1990)

The Epileptic Focus versus the Pathological Focus

A. Rougier

Clinique Universitaire de Neurochirurgie, Group hospitalier Pellegrin, Bordeaux, France

Summary

The relationship between the epileptic focus and the causative lesion has been studied to define when resection of the epileptic focus may be associated with removal of the lesion. Experimental studies reveal that a maturation phase is followed by a progressive autonomisation of the epileptic focus from the induced lesion. Thereafter there may be a progressive course to such experimental epilepsy with spreading of the epileptic lesions and possible creation of a secondary focus. However, in man the findings are less coherent and sometimes contradictory. In long term intractable epilepsy, simple removal of the causative lesion can be effective. A hypothesis is proposed to explain these phenomena – the hypothesis of systemic focal epileptogenesis. It is further proposed that endogenous excitotoxic agents such an quinolinic acid may have effects on the epileptic focus only when the causative lesion is present producing a subtle modification of the blood brain barrier.

Keywords: Epilepsy; epileptic focus; causative lesion.

Introduction

The epileptic focus can be defined as the cortical region in which seizures originate and whose removal brings about their disappearance. The delineation of this cortical zone is derived from two types of data. Clearly the manifestation of epileptic hyperexcitability such as ictal behaviour, interictal spikes, and ictal discharges provide considerable information. However, interictal signs of focal functional deficits such as non-epileptiform EEG abnormalities and PET-identified hypometabolism also help to define the epileptic focus. Histological examination of the resected cortex often reveals lesions directly connected with the epileptic phenomenon: mesial temporal sclerosis, although morphologically non-specific, is observed in temporal lobe epilepsy with a statistically high incidence[23]. Morphometric studies have shown that neuronal loss and dendritic abnormalities are correlated with sustained seizures. Because the epileptic focus is characterised not only by specific hyperexcitability but also by func-

tional deficits and structural involvement, the term epileptic lesion is often used in place of epileptic focus. Consequently, an ambiguity can arise between this entity and another type of lesion which induces the epileptic focus in its vicinity. These lesions are various and include: scars of diverse origin, hamartomas, slowly growing tumors and cryptic vascular malformations. Until recently many of these lesions were revealed only at operation. It was recommended that, when such a lesion was disclosed during an operation, the electrically firing cortex in the vicinty (epileptic focus) should also be excised in addition to the lesion[21]. Nowadays, CT scanning and MRI permit the discovery of such lesions prior to surgery. This advance knowledge of specific lesions responsible for epilepsy has modified our concepts of diagnosis and management. This situation requires re-examination of the question of the relationships between the epilepsy-inducing lesion and the epileptic focus itself: to relieve the epileptic condition, should the removal of the epileptic focus be associated with the resection of the epilepsy-inducing lesion? To try to answer this question, the relationships between the epilepsy-inducing lesion and the epileptic focus will be explored at different stages of their evolution.

Experimental Findings

Whatever the method used to produce a chronic epileptic focus, seizures occur after a maturation phase lasting several days or weeks. During this phase structural lesions evolve early in the perilesional cortex. For example, progressive reduction of thickness and selective loss of the neurons of layers 2 and 4 are observed around alumina cream granulomas[32]. Moreover functional modifications appear before any clinical or electric epileptic manifestations as decreasing densitiy of noradrenergic containing terminals in a cobalt focus[4]. The maturation phase is divided into two stages in alumina-cream inducing focal motor epilepsy[32]. The latent pre-mature stage is characterised by EEG spikes during sleep and the latent mature stage by an increasing number of spikes also present during wakefulness. If the A-C granuloma is excised at the latent pre-mature stage no clinical convulsion will occur. However, the anti-epileptic effect of A-C granuloma excision during the latent mature stage is small. Thus, the mechanisms underlying epilepsy are induced early and can evolve independently of the initial lesion. This phenomenon is found even in kindling. Initial applications of the stimulations may or may not provoke local epileptiform after-discharges. Provided that stimulation parameters exceed some threshold value a local after-discharge will appear after several repetitions. With later trials after-discharges increase, and stimulations elicit 5 classes of seizures. Spontaneous generalised motor seizures are infrequently observed. Kindled seizure susceptibility persists for up to twelve months in cats. If stimulation is discontinued before the change becomes permanent, the thresholds required to elicit after-discharges re-increase gradually after a lapse of time.

In the convulsive stage the resection of the A-C-granuloma does not significantly reduce the seizure frequency. In order to eliminate epilepsy a second resection of the surrounding cortex judged epileptic by ECoG must be performed[15]. No alumina was encountered in this tissue after the resection of the granuloma and the immediate surrounding involved cortex. Thus, it appears that the ensuing epilepsy in animals in which the inciting alumina had been excised would represent epilepsy in the absence of alumina[32]. In kindled primates, convulsive responses can be obtained at brain sites distant from the kindled amygdala. Wada observed that recurrent convulsive seizures continued following amygdalectomy. This finding suggests that kindling seizure susceptibility and the availability of its neurocircuits for seizure generation do not depend on the kindled focus[33].

Much data suggests that experimental focal epilepsy can be a progressive and not a self-limiting entity. Sustained seizures induce brain damage and continuing clinical seizures are associated with continuing neuronal damage. They are specific lesions regardless of the inductory method employed and can be interpreted as a seizure-mediated phenomenon. These lesions closely resemble those produced by excitotoxic mech-

anism mediated by glutamate or aspartate[24]. Moreover, the excitatory neurotransmitters play a role in the mechanisms of epilepsy. Thus it may be that the occurrence of seizures facilitates other new seizures, and the extension of the epileptic focus. Other evidence for a progressive course is the development of an independent secondary epileptic focus. Non-primate experiments have clearly demonstrated that a unilateral epileptogenic lesion, if left untreated, will sooner or later give rise to epileptiform discharges in the contralateral region. This phenomenon is not clearly defined in the primate. Initially the secondary focus is dependent and immediately disappears if the primary focus is excised. During an intermediate stage, independent inter-ictal spikes are observed. After excision of the primary focus the spikes subside over a period of weeks. In the independent stage, the secondary focus becomes autonomous and the excision of the primary has no effect[18].

From such experimental findings, epilepsy can be interpreted as a phenomenon that, before becoming stable, requires a maturation phase of variable duration. Once established, the epilepsy becomes independent of the lesion that created it and may undergo a progressive course: spread of the epileptic lesion and possible creation of a secondary focus.

Clinical Findings

We have attempted to determine if the previous data are relevant to human epilepsy. As with an experimental chronic focus, epilepsy in man requires a maturation phase. Sometimes the data of origin of the epilepsy-inducing lesion is well-defined as in post-traumatic or post-ischaemic epilepsy. The lapse between the initial incident and the appearance of epilepsy is variable but sometimes very long. If the removal of the lesion is performed during this phase, we would suppose that this surgical procedure would be sufficient to avoid the creation of a chronic and perhaps intractable epilepsy. Such a question is raised when, after a first seizure, a haemangioma calcifians is discovered. This benign variant of cavernous angioma never involves haemorrhage[26]. The only risk is the creation of intractable epilepsy. A wait-and-see attitude is recommended by some[8]. Others advocate removal because of its epileptogenic potential. In this situation surgical ablation is effective[28] given the assumption that the epileptic focus is merely in its maturation phase and not yet independent of the epilepsy inducing lesion. When medically uncontrolled seizures exist for a long time, epilepsy surgery is sometimes recommended. A debate still persists over the appropriate surgical procedure when a specific pathological focus is at the origin of the epilepsy.

Persistence of seizures after the removal of the causative lesion has been noticed in various pathologies. Drake et al.[9] reported sixteen patients with temporal lobe epilepsy and mass lesions. Nine of them had concomittant mesial temporal sclerosis. Such an association supports the suggestion that hippocampal changes are secondary to repetitive seizures and may act as a source of subsequent seizure activity. Consequently the surgical management of these cases requires that both problems be addressed: because the epileptic activity arises not in the mass lesion but in the surrounding cortex, simple excision of the lesion may not eradicate the epilepsy. To define the cortical area which may be resected, the topographical relations between slowly growing tumors and the epileptic area have been explored by stereo-EEG. These studies point out that inter-ictal spikes and ictal discharges can involve cortical zones distant from the tumor[2, 19, 25]. Likewise surgery of arterio-venous malformations does not reduce epilepsy significantly[6, 20]. But surgery aims at eliminating the shunt and not at removing the intense reactive gliosis surrounding the malformation that is present sometimes.

In contrast, the incidence and severity of seizures may be reduced by excision of an epilepsy-inducing lesion without resection of the epileptic focus that has been observed on several occasions. Falconer et al.[11] have reported the disappearance of temporal seizures after removal of lesions that happened to be located a certain distance from the supposed epileptic focus. Spencer et al.[29] found that all patients treated for mass lesions regardless of the type of method of treatment had more than a 75% reduction in seizure frequency. But patients treated specifically with surgery for the epilepsy had more than 95% reduction in seizures frequency. Dramatic control of intractable seizures was obtained in three cases with secondary bilateral synchronization by surgical removal of circumscribed unilateral pial angiomatosis[5]. But in one case the underlying calcified cortex was also removed and, in the second case, an occipital lobectomy was performed. Thus the resection of a part or the whole of the epileptic cortex surrounding the lesion was associated with the removal of the epilepsy inducing lesion. Seven cases of thrombosed arterio-venous malformations, presenting with intractable seizure disorderd, became seizure-free after simple excision[34]. Such observations have been made with cavernomas[17, 28], calcified choristomas[1] and childhood brain tumors[3, 12].

Results

We report the surgical results of 71 cases operated on for long-term epilepsy refractory to medication according to the surgical procedure performed. In 60 cases a cortical resection of the epileptic focus was carried out with, obviously, excision of the epilepsy-inducing lesion when one existed (SURGERY 1). In 11 cases a simple removal of the lesion was performed because the epileptic focus was located in highly functional areas (SURGERY 2). No difference exists between the length of the epilepsy and the age of the patients (Table 1). The epilepsy-inducing lesions were of various types but they were equally distributed in the two groups (Table 2). Even though the second group is small, no statistically significant difference was found between the two types of surgical approach (Fig. 1). The same findings have been observed previously in chronic focal epilepsy induced by astrocytomas[14]. A comparative follow-up study of 48 cases with epileptic focus resection and 523 cases with standard removal of the tumor has been performed. In the group treated

Table 1. *Presurgical Characteristics*

		SURGERY I 60	SURGERY II 11
Time between seizure onset and surgery	mean	14 years	15 years
Age at surgery	mean	28 years	30 years

Table 2. *Histological Examination*

Pathology	SURGERY I	SURGERY II
Pathological focus		
cavernomas	4	6
dermoid cysts	2	–
glial hamartomas	1	1
indolent gliomas	2	2
xanthoastrocytomas	1	1
post-traumatic scars	4	–
post-ischaemic scars	2	1
Epileptic focus		
non-specific gliosis	16	
mesial temporal sclerosis	10	
other minimal lesions	8	
no convincing abnormality	7	

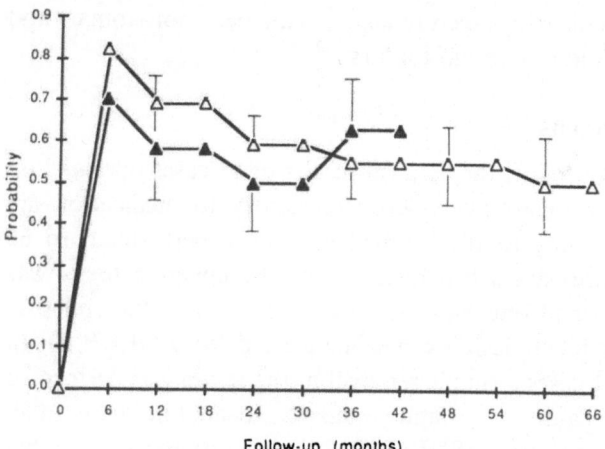

Fig. 1. Probability of becoming seizure-free according to surgical procedure. △ Surgery I, ▲ surgery II (see text). Statistical study by Markov's method: no significant difference was found

by epilepsy surgery 46% were seizure-free and 6% still experienced seizures. In the group treated by tumors removal, only 31% were seizure-free but persistence of seizures were noted in a similar number (8%). The difference in the number of seizure-free patients between the two groups is accounted for by the recurrence of tumor and death.

Discussion

Why does a chronic epileptic focus require the presence of a lesion to be active? Firstly, it is known that the presence of a lesion is a favorable prognostic factor for successful epilepsy surgery. An epileptic genetic predisposition is usually missing whereas a low epileptic threshold can exist in some cases of symptomatic focal epilepsy that do not have a precise origin. For example, amygdala kindling is obtained easier and faster in photosensitive papio than in the rhesus monkey[33]. To explain why the activity of the potential epileptic focus is dependent on the presence of a causative lesion the hypothesis of systemic focal epileptogenesis is relevant. Remler et al.[22] first induced a focal blood brain barrier lesion by irradiation. Systemic administration of a convulsant drug that cannot cross the normal BBB then produced focal epilepsy in the region where the BBB is defective. Thus a chronic focus is created by systemic administration of repeated injections of bicuculline. Many lesions that result in chronic focal epilepsy are associated with focal BBB breakdown. Furthermore, convulsants that cannot cross the normal BBB are common in normal serum. The excitotoxic and convulsant quinolinic acid (QUIN) could serve as an endogenous

epileptogen[27]. Thus pathological intracerebral accumulation of QUIN could result from the persistence of BBB breakdown due to the persistence of the lesion. We will have to prove that the BBB is modified around lesions without CAT enhancement. Subtle changes have been observed suggesting a possible transendothelial transport in experimental astrocytomas[7]. Moreover we have studied the activity of quinilinic-phosphoribosyl transferase (QPRTase) which is the first degradative enzyme of QUIN in epileptic foci. A specific reduction of QPRTase activity was observed in tissue primarily involved in the epileptic discharge and may contribute to the accummulation of QUIN and the maintenance of an active epileptic focus[13]. We can postulate that the removal of the epilepsy inducing lesion leads to the disappearane of the systemic epileptogenesis by closing the BBB. Consequently, the focal epileptic focus, isolated from the excitotoxic input, becomes extinct.

Conclusion

Experimental data are relatively consistent in suggesting that there is progressive autonomization of the epileptic focus from the epilepsy inducing lesion. Observations made in man are less coherent, sometimes even contradictory. We can observe that structural lesions and functional abnormalities do not have the same enlargement[16, 31]. Epilepsy-inducing lesions differ from electrically defined epileptic foci. The inter-ictal spiking cortical area is wider than the region primarily involved by the ictal discharges[30]. The PET-defined hypometabolism spreads to other regions than the electrically defined epileptic focus[10]. Moreover, secondary epileptogenesis in man, inferring that epilepsy could have a progressive course, is difficult to prove although Morrel provides some convincing arguments[18]. When we have to choose the surgical procedure for intractable epilepsy induced by a specific lesion, the relative merits of each approach must be discussed given the efficacy of simply removing the epilepsy-inducing lesion in many patients.

References

1. Averback P (1977) Epileptogenic mineralization: pathological variants with good prognosis. Ann Neurol 2: 332–335
2. Bancaud J, Talairach J, Geier S, Scarabin JM (1973) EEG at SEEG dans les tumeurs cerebrales et l'epilepsie. Edifor Paris, 346 pp
3. Blume W, Girvin J, Kaufmann J (1982) Childhood brain tumours presenting as chronic uncontrolled focal seizure disorders. Ann Neurol 12: 538–541

4. Chauvel P, Trottier S (1986) Role of noradrenergic ascending system in extinction of epileptic phenomena. In: Delgado-Escueta A V *et al* (eds) Advances in neurology, vol 44. Raven Press, New York, pp 475–487

5. Chevrie J, Specola N, Aicardi J (1988) Secondary bilateral synchrony in unilateral pial angiomatosis: successful surgical treatment. J Neurol Neurosurg Psychiatry 51: 663–670

6. Crawford PM, West CR, Shaw DM, Chadwick DW (1986) Cerebral arterio-venous malformations and epilepsy: factors in the development of epilepsy. Epilepsia 27: 270–275

7. Deane BR, Lantos PL (1981) The vasculature of experimental brain tumours: a quantitative assessment of morphological abnormalities. J Neurol Sci 49: 67–77

8. Ditullio M, Stern E (1979) Hemangioma calcifians: case report of an intraparenchymatous calcified vascular hematoma with epileptogenic potential. J Neurosurg 50: 110–114

9. Drake J, Hoffmann HJ, Kovakashi J, Hwang P, Becker LE (1987) Surgical management of children with temporal lobe epilepsy and mass lesions. Neurosurgery 21: 792–796

10. Engel J, Brown J, Kuhl D, Phleps D, Mazziotta J, Crandall P (1982) Pathological findings underlying focal temporal hypometabolism in partial epilepsy. Ann Neurol 12: 518–528

11. Falconer MA, Driver MV, Serafetinides EA (1962) Temporal lobe epilepsy due to distant lesions: two cases relieved by operation. Brain 85: 521–534

12. Farwell J, Stuntz T (1984) Fronto-parietal astrocytoma causing absence seizures and bilaterally syncronous epileptiform discharges. Epilepsia 25: 695–698

13. Feldblum S, Rougier A, Loiseau H, Loiseau P, Cohadon F, Morselli PL, Lloyd KG (1988) Quinolinic-Phosphoribosyl transferase activity is decreased in epileptic human brain tissue. Epilepsia 29: 523–529

14. Gonzales D, Elvidge A (1962) On the occurrence of epilepsy caused by astrocytoma of the cerebral hemispheres. J Neurosurg 19: 470–482

15. Harris AB, Lockard JS (1981) Absence of seizures or mirror foci in experimental epilepsy after excision of alumina and astrogliotic scar. Epilepsia 22: 107–122

16. Holmes M, Kelly K, Theodore W (1988) Complex partial seizures. Correlation of clinical and metabolic features. Arch Neurol 45: 1191–1193

17. Jabbari B, Huott A, Dichiro G, Martins A, Coker S (1978) Surgically correctable lesions detected by CT in 143 patients with chronic epilepsy. Surg Neurol 10: 319–322

18. Morrel F (1985) Secondary epileptogenesis in man. Arch Neurol 42: 318–335

19. Munari C, Musolino A, Blond S, Brunet P, Giallonardo AT, Chodkiewicz JP, Bancaud J (1986) Stero-EEG exploration in patients with intractable epilepsy: topographic relations between a lesion and epileptogenic areas. In: Schmidt D *et al* (eds) Intractable epilepsy. Raven Press, New York

20. Murphy MJ (1985) Long-term follow-up of seizures associated with arteriovenous malformations. Results of therapy. Arch Neurol 42: 477–479

21. Rasmussen T (1975) Surgery of epilepsy associated with brain tumours. In: Purpura DP *et al* (eds) Advances in neurology, vol 8. Raven Press, New York, pp 227–239

22. Remler M, Marcussen W (1986) Systemic focal epileptogenesis. Epilepsia 27: 35–42

23. Robitaille Y (1987) Neuropathological findings in epileptic foci. In: Wieser HG *et al* (eds) The epileptic focus. John Libbey, London, pp 95–112

24. Olney JW, Collins RC, Sloviter RS (1986) Excitotoxic mechanisms of epileptic brain damage. In: Delgado-Escueta L *et al.* (eds) Advances in neurology, vol 44. Raven Press, New York, pp 857–877

25. Rossi GF, Colicchio G, Gentilomo A, Pola P, Scerrati M (1980) The stereoencephalogram as a means to locate epileptogenic lesions. In: Canger *et al* (eds) Advances in epileptology. XIth Epilepsy International Symposium. Raven Press, New York, pp 91–101

26. Sanchis Fargueta J, Iranzo R, Garcia M Jorda M (1980) Hemangioma cacifians. A benign epileptogenic lesion. Surg Neurol 15: 66–70

27. Schwarcz R, Whetsell WO Jr, Mangano RM (1983) Quinolinic acid: an endogenous metabolite that produces axon-sparing lesions in rat brain. Science 219: 316–318

28. Simard JM, Garcia-Bengochea F, Ballinger W, Mickle J, Quisling R (1986) Cavernous angioma: a review of 126 collected and 12 new clinical cases. Neurosurgery 18: 162–171

29. Spencer DD, Spencer SS, Mattson RH, Williamson PD (1984) Intracerebral masses in patients with intractable partial epilepsy. Neurology 34: 432–436

30. Talairach J, Bancaud J (1966) Lesion, irritative zone and epileptogenic focus. Confin Neurol 27: 91–94

31. Theodore W, Holmes M, Dorwart R, Peorter R, Di Chiro G, Sato S, Rose D (1986) Complex partial seizures: cerebral structure and cerebral function. Epilepsia 27: 576–582

32. Velasco M, Velasco F, Pacheco MT, Azpeita E, Saldivar L, Estrada-Villaneuva F (1984) Alumin-cream induced focal motor epilepsy in cats, Part 5. Excision and transplant of the epileptogenic granuloma. Epilepsia 25: 752–758

33. Wada JA, Naquet R (1986) Experimental model in a primate predisposed to epilepsy. In: Schmidt D *et al* (eds) Intractable epilepsy. Raven Press, New York, pp 39–59

34. Wharen RE, Scheithauer BW, Laws ER (1982) Thrombosed arteriovenous malformations of the brain. An important entity in the differential diagnosis of intractable focal seizure disorders. J Neurosurg 57: 520–526

Correspondence: A. Rougier, M.D., Clinique Universitaire de Neurochirurgie, Group hospitalier Pellegrin, F-33076 Bordeaux Cedex, France.

Acta Neurochirurgica, Suppl. 50, 6–13 (1990)

Membrane Electrophysiology of Epileptiform Activity in the Hippocampus

H. V. Wheal

Department of Neurophysiology, University of Southampton, U.K.

Summary

Several different types of Na^+ and Ca^{++} channels in the membrane of neurones provide the driving force for excitation. Whilst some of these may be activated by the transmembrane voltage or ionic concentrations, others are mediated by neurotransmitters and neuromodulators. The role of these ionic mechanisms in epileptiform activity are discussed, with particular reference to the involvement of the NMDA receptor mediated channel.

Multiple K^+ and Cl^- mediated mechanisms provide the stabilizing influence on the electrophysiological behavior of the cells. Loss or reduction in activity of one or more of these conductances may lead to the expression of epileptiform activity. The role of the extracellular concentration of K^+ in burst firing of populations of cells is discussed.

The examples are primarily chosen from studies of the hippocampus in animal models of epilepsy, however wherever possible experiments on human tissue have been discussed. These studies on the membrane and synaptic mechanisms that contribute to epileptiform activity provide us with the necessary insights to allow the development of new methods for controlling such activity.

Keywords: Epilepsy; hippocampus; synaptic function; ionic channels; excitatory amino acids; GABA inhibition; review.

Introduction

There are many processes, both cellular and synaptic that may generate epileptiform activity in groups of neurons in the mammalian central nervous system. Bursting activity or seizure discharges have for a long time been perceived as a balance between mechanisms that cause excitation and those that cause inhibition of the cell. At a cellular level we first have to discuss the ionic processes in the membrane.

Ionic Processes in the Membrane

The basic foundations are the differences in the concentration of the ions Na^+, K^+, Cl^- and Ca^{++} between the inside and the outside of the cells, and the processes that control them. Superimposed on these are the fundamental components, or building blocks such as the voltage and neurotransmitter gated ionic channels. It was Hodgkin and Huxley[40] who pioneered the work on the voltage-dependent ionic conductances by describing the mechanisms responsible for the action potential in the squid axon. The depolarizing phase of the action potential was caused by a transient Na^+ conductance, whilst the repolarization was facilitated by a slower K^+ conductance. These components are still used in models of neuronal activity today. However due to the advent of mammalian *in vitro* preparations[46, 23] as well as significant development in electronics, many different types of ionic channel and their associated currents have now been discovered[39, 55, 69].

An increase in the permeability or conductance of the membrane to a specific ion allows a current to flow through the membrane carried by that ion. The consequence of this inward or outward current flow is a change in the potential across that membrane. If the membrane potential is restrained, or voltage clamped, the current flowing across the membrane can be measured. In this way it is possible to measure the current flowing at different membrane potentials, ie those positive to (depolarizing) or negative to (hyperpolarising) the resting potential (usually -60 to -70 mV). If neurotransmitter receptors are activated, they may produce an increase or decrease in the conductance of the membrane which can be measured as a change in the current under voltage clamp conditions.

Although a variety of *in vitro* models of epilepsy have been used for experiments on individual cells, the examples for this article will primarily be taken from work on the hippocampus (see 73). This nucleus has attracted considerable attention not only due to its role in learning and memory but also its involvement with the genesis of focal epileptic activity of the temporal lobes and its accessibility for *in vitro* study.

Na^+ and Ca^{++} Mechanisms

In addition to the transient Na^+ conductance already mentioned, pyramidal cells in the hippocampus also possess a slow, persistent or non-inactivating Na^+ conductance[11,57]. As in cells from the cerebellum and cortex, this is a subthreshold conductance which does not generate action potentials, but does trigger high threshold, fast Na^+ spikes[56,77]. It can also generate long plateau potentials which may change the firing frequency of the cell or modify excitatory postsynaptic synaptic potentials (EPSPs). Thus although it appears to be a subtle conductance, it may contribute to the generation and maintenance of an epileptiform burst.

Calcium conductances also contribute significantly to the excitability and behavior of pyramidal cells. One subgroup of pyramidal cells, those in CA 3 region of the hippocampus, spontaneously fire bursts of action potentials. Slow, high threshold Ca^{++} conductances may underlie these pacemaker potentials[43,15]. In the dendrites of these cells, Ca^{++} − dependent action potentials have been reported[71]. There is a second Ca^{++} conductance that may also contribute to the oscillatory behavior of the cells, particularly if the membrane potential is hyperpolarised in between bursts. This event is triggered at a very low threshold and is rapidly inactivated at potentials just above the resting potential of the pyramidal cells[33]. In this case, hyperpolarising the cell below its resting potential primes it for the next transient Ca^{++} depolarization.

In more recent voltage clamp studies on cultured dorsal root ganglion cells, Fox, Nowycky and Tsien[26] have characterised the pharmacological properties of three calcium currents (L, T and N). They have separated these currents using selective Ca^{++} channel blockers. Cadmium ions (20–50 uM) were found to block both N and L currents, whilst nickel ions (100 uM) strongly reduced the T current but left the N and L currents intact. Nifedipine (10 uM), the dihydropyridine antagonist, was found to specifically inhibit the L current. The N and L type currents are thought to correspond to the high threshold Ca^{++} conductances that may contribute to dendritic spiking, whereas the T-current may be activated at a low voltage. Of particular interest at the moment is the site of these Ca^{++} channels in the cell, ie their dendritic or somatic location and their pre- or postsynaptic physiological function.

Ca^{++} Fluxes in Epileptic Tissue

One of the limitations of recording ionic conductances or currents from individual neurons is that one has no idea what the collective consequence of that conductance is on a population of cells. During interictal or ictal bursting, large numbers of neurons are involved. Measuring the concentration of the calcium in the extracellular environment ($[Ca^{++}]o$) using ion sensitive electrodes is a particularly good method of monitoring the involvement of that ion in population responses. Using this technique, Heinemann et al.[37] have compared the changes in $[Ca^{++}]o$ in a wide range of chronic and acute models of epilepsy. In virtually all these models there was a greater reduction in $[Ca^{++}]o$ following epileptiform activity when compared with normal synaptic activity. Some models also showed a change in the spatial distribution of the Ca^{++} fluxes. For example in slices of normal hippocampus the maximum decrease in $[Ca^{++}]o$ following orthodromic activation was close to the pyramidal cell layer, which contains the soma of the pyramidal cells. In slices of hippocampus taken from kindled rats that showed tonic-clonic convulsions, the areas of maximum Ca^{++} flux had spread to the dendritic synaptic areas in the stratum radiatum[85].

Although as described above, low $[Ca^{++}]o$ is a consequence of epileptiform acitivity, low $[Ca^{++}]o$ in the hippocampus itself also induces hyperexcitability and spontaneous epileptiform activity[88]. One can think of this as a process of amplification of epileptiform activity in the tissue that requires other ionic or synaptic mechanisms to control it. Several possible membrane and synaptic mechanisms may be affected by low $[Ca^{++}]o$, such as a failure of calcium dependent potassium conductances or a block of inhibitory synapses have been suggested[44]. Both these mechanism would lead to the hyperexcitability of a population of cells and are discussed in more detail below.

Role of NMDA Receptors in Epileptiform Activity

A unilateral intracerebroventricular (ICV) injection of kainic acid (KA) results in a specific loss of ipsilateral CA 3/4 pyramidal cells in the hippocampus in rats[65,53], a lesion which is similar to that seen in human temporal lobe epilepsy[62]. This KA-induced lesion is associated with the appearance of multiple population spikes, or epileptiform bursting activity, as shown in Fig. 3[54]. Expression of the epileptiform activity in this model appears to involve changes in both inhibitory and excitatory synaptic function[87].

N-methyl-D-aspartate (NMDA)-receptors are a subclass of excitatory amino acid receptor found in the mammalian central nervous system (see[23]). In the hip-

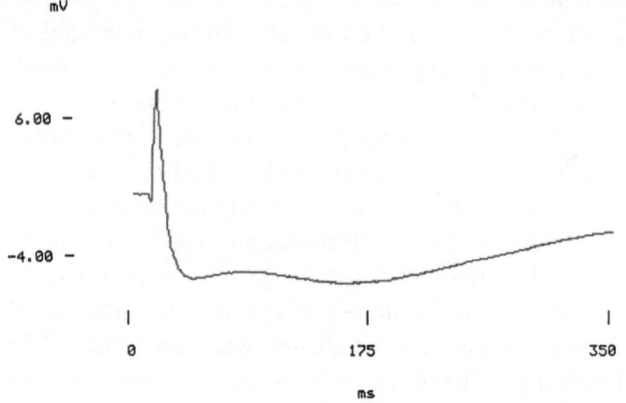

Fig. 1. Intracellular recording from a control pyramidal cell in the hippocampus, resting potential $-64\,mV$. This recording illustrates the fast Na^+/K^+ mediated EPSP followed by the early and late IPSPs mediated by Cl^- and K^+ respectively. Note the short duration of the excitatory event compared with the much longer duration of the inhibition

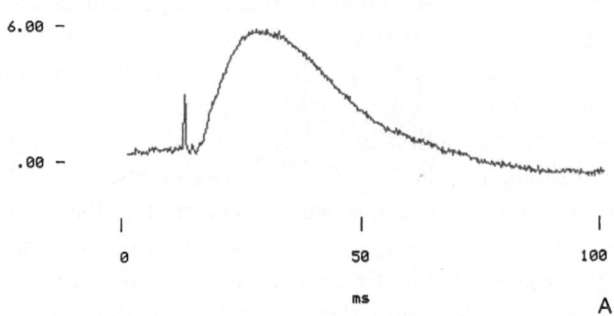

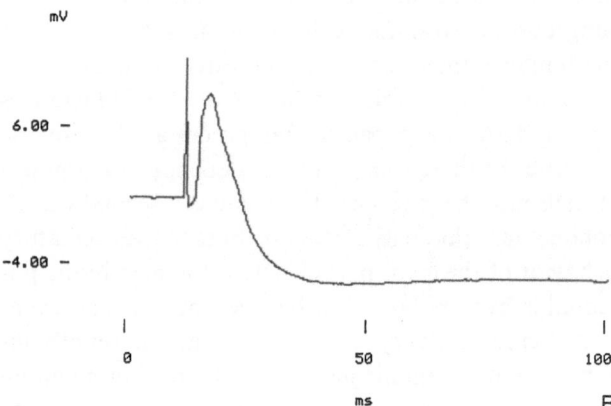

Fig. 2. Intracellular recording from CA 1 pyramidal cells in the KA lesioned hippocampus (A) and control unlesioned hippocampus (B). A) This EPSP underlies the epileptiform activity and is mediated by NMDA-receptors. Both Na^+ and Ca^{++} are thought to contribute to the shape of this EPSP. Note also the virtual absence of any synaptic inhibition. B) Control EPSP mediated by non-NMDA-receptors. The fast decay phase of this EPSP is caused by the early $GABA_A$ receptor mediated IPSP (from Turner and Wheal, 1988)

pocampus, evoked postsynaptic excitatory potentials (EPSPs) recorded from CA 1 pyramidal cells in the hippocampus, shown in Fig. 1 and 2, are normally insensitive to NMDA-receptor antagonists[49, 20]. However, specific NMDA-receptor antagonists including D-amino-5-phosphonovalerate (D-APV) have been found to be effective in blocking seizures in several *in vivo* model of epilepsy[87, 22, 63]. D-APV also blocks the epileptiform activity in hippocampus *in vitro* induced by low $[Mg^{++}]o$ which removes a voltage-dependent Mg^{++} block of the channel linked to the NMDA-receptor complex[38, 67].

D-APV was found to be a potent specific blocker of the EPSP underlying epileptiform bursts of action potentials in recordings from pyramidal cells in the KA lesioned hippocampus *in vitro*[6, 7]. In studies by Turner and Wheal[84], the subthreshold EPSP's shown in Figure 2 were almost entirely mediated by NMDA-receptors. The mechanism by which the subtype of exctatory amino acid receptor switches from non-NMDA to NMDA is of particular interest and may hold the key to novel treatments of temporal lobe epilepsy. This approach is especially promising since NMDA-receptor mediated seizure discharge has been recorded *in vitro* from human epileptogenic neocortical tissue removed during temporal lobectomy[8, 9].

In addition to the voltage dependence of the Mg^{++} block of the channel in the NMDA-receptor complex, the other property that is important to paroxysmal activity is its ability to carry Ca^{++} through the membrane[67]. Repetitive application of NMDA to hip-

pocampal neurones leads to a long lasting increase in the intracellular $[Ca^{++}]$[21]. The channels associated with the other excitatory amino acids are more selective for Na^+ and K^+ and thus little Ca^{++} enters the pyramidal cells in the hippocampus during normal synaptic activity.

K^+ Mediated Mechanisms

Potassium channels play an important role in regulating the excitability of neurons. The $[K^+]$ gradient causes increased K^+ conductance to produce outward currents that hyperpolarise the cell and thus inhibit activity of the cell. If the three types of Ca^{++} current are confusing then the number and diversity of K^+

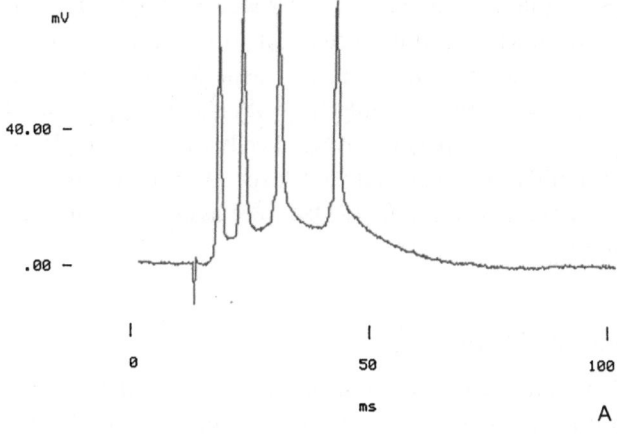

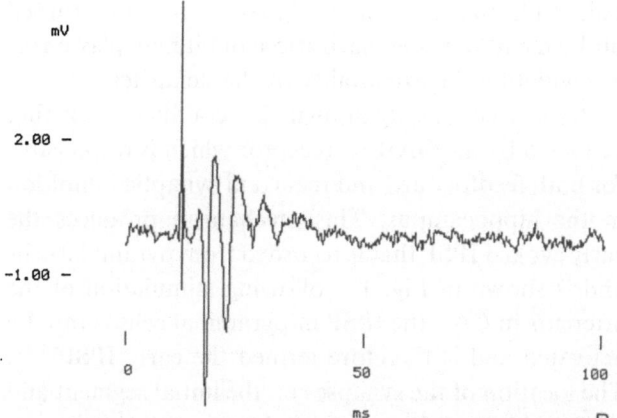

Fig. 3. Stimulus evoked epileptiform bursts of action potentials recorded from the CA 1 cells in the KA lesioned rat hippocampus. A) Intracellular recording from a single pyramidal cell in that region. Note the long lasting EPSP underlying the burst and the absence of any synaptic inhibition. Under control conditions only a single action potential can be evoked from these cells. B) Extracellularly recorded epileptiform field burst of population spikes. The amplitudes of these spikes reflect the number of cells that are contributing to the event. A large number of pyramidal cells fire synchronously to generate this activity (from Ashwood, Lancaster and Wheal, 1986)

currents will at first seem incomprehensible. However, it is their diversity that offers the pharmacologist and molecular biologist the most sophisticated tools for controlling neural activity. Potassium channels are regulated by many different mechanisms, including membrane voltage, drugs, neurotransmitters, toxins and second messengers. They are also found in many different types of cell, and this ubiquity has recently been reviewed by Rudy[69].

The K^+ current described by Hodgkin and Huxley is termed a delayed rectifier (I_K). Such a current has been found in cultured hippocampal neurones[74] and is probably partly responsible for the repolarisation of the action potential. It is also partly responsible for the duration of the refractory period of the cell, thus setting an upper limit on its firing frequency. Another voltage dependent K^+ current is the transient "I_A" current, which in CA 3 and CA 1 pyramidal cells is active at the resting membrane potential (-60 mV)[28, 36]. The I_A current can be identified by its sensitivity to 4-aminopyridine (4-AP). This current may control the latency of burst onset or regulate the initial firing frequency of the cell[78].

In addition to the voltage dependent K^+ mechanisms that contribute to the repolarization of action potentials in pyramidal cells, there are Ca^{++} mediated processes[79, 52]. The fast Ca^{++} dependent K^+ conductance (I_C) can be blocked by tetraethylammonium (TEA) or a component of scorpion venom, known as charybdotoxin (CTX). Once this K^+ current is blocked the duration of the action potential increases, and there is also an increase in the initial firing frequency of the burst action potentials induced by depolarizing current. Following such a burst in CA 1 pyramidal cells there is a long lasting afterhyperpolarization (AHP) that is also mediated by a Ca^+ dependent K^+ current (I_{AHP})[41, 29, 60, 54, 51]. The slow AHP can be blocked by Ca^+ chelators eg. EGTA and BAPTA[72, 54, 60, 52, 82]. By using such blockers the slow AHP was shown to cause adaptation of the firing frequency of the cell. During an extended burst (> 500 ms) the firing frequency of the action potentials gradually reduces, or adapts. Furthermore, the slow AHP conductance has been shown to be modulated by a large number of neurotransmitter agents, acetylcholine, nordrenaline, histamine, dopamine, adenosine and 5HT[12, 59, 60, 16, 17, 18, 30, 31, 61, 68, 3, 19].

The M-current (I_M) is another K^+ current that is voltage and time-dependent as well as being modulated by acetylcholine, through muscarinic receptors[14, 35]. This mechanism was originally identified in frog sympathetic neurones and has subsequently been characterized in mammalian and human neocortical neurones[34]. This K^+ current is activated at potentials at or above the resting membrane potential (approx -60 mV). It is blocked by muscarinic agonists, which leads to a depolarization and increase in excitability and firing frequency of the cells[34]. This is some of the evidence that suggests that the properties of human hippocampal and cortical neurones are similar to those of other mammals.

Failure of Slow K^+ Conductances in the KA Model

At least two slow K^+ conductances appear to be reduced or lost in the KA model of temporal lobe epilepsy. The first of these is the AHP described above, that is mediated by a slow Ca^{++} activated K^+ conductance. In intracellular recordings from surviving pyramidal neurones in the lesioned hippocampus there was a loss of the AHP following a burst of action potentials as well as spike frequency accomodation. This was associated with an increase in the firing frequency of the cells, and their general excitability. In the absence of the slow AHP, the epileptiform burst may be terminated by a more transient Ca^{++} activated K^+ conductance[5, 2].

Another slow K^+ conductance normally found in pyramidal cells is that activated by $GABA_B$ receptors[13, 1, 54, 66, 27]. This is a small conductance that hyperpolarised the cell, forming the late IPSP that follows the orthodromic activation shown in Fig. 1. The late IPSP is bicuculline insensitive but has recently been blocked by the baclofen antagonist, phaclofen[47, 75, 24]. This feedforward inhibitory mechanism also appears to be reduced or absent in the KA lesioned hippocampus[5].

We do not yet understand the mechanism of loss of these two slow K^+ conductances that contribute to the epileptiform activity. One suggestion is that there could be a change in the intracellular secondary messenger mechanisms that may be responsible for their activation. However, modulation of these long lasting hyperpolarizations is critically important in stabilizing membrane and synaptic mechanisms.

Extracellular K^+ Ion Concentration

Experiments on either chronic or acute models of epilepsy have shown that the extracellular potassium ion concentration $[K^+]o$ rises dramatically during interictal and prolonged ictal like discharges[76, 37, 70, 80, 81]. Selective ion sensitive electrodes have been used in experiments to monitor baseline $[K^+]o$ levels (5 or 6 mM), which rose to values of 10–20 mM during the epileptiform discharges. This increase in $[K^+]o$ will have a complex action on the cell and synapses, however the most obvious effect is a decrease in the driving force of the K^+ conductances leading to an increase in the frequency of interictal activity or maintained discharge of the neurones. During postictal depression, the $[K^+]o$ may be below its resting level, which contrasts with the very high $[K^+]o$ found in Leao's or spreading depression[76, 45].

It has been suggested that the increase in $[K^+]o$ follows the onset of interictal bursting or seizure activity, and therefore is not responsible for generating the epileptic acitivtiy[76]. However, high $[K^+]o$ (> 6.5 mM) in the absence of any other convulsant has been shown to spontaneously generate interictal and seizure activity in populations of cells in hippocampal slices[37, 70, 81]. Furthermore intracellular recording from pyramidal cells showed that large excitatory paroxysmal depolarizing shifts (PDS) were associated with this activity[70].

Role of Cl^- Ions

A voltage-sensitive chloride current ($I_{Cl(V)}$) has been found in pyramidal cells in the hippocampus[58]. This current is active at the resting potential of the cell and therefore contributes to the resting conductance of the cell. Madison et al.[58] reported that the current is turned off by membrane depolarization and it may play a role in regulating the excitability of the dendrites.

A more commonly known Cl^- conductance is that activated by the $GABA_A$-receptor which is responsible for both feedforward and recurrent synaptic inhibition in the hippocampus. This mechanism produces the early evoked IPSP that is recorded from pyramidal cells and is shown in Fig. 1. Following stimulation of the afferents in CA 1 the IPSP in pyramidal cells is rapidly activated and is therefore termed the early IPSP[64, 83]. The location of the synapses on the initial segment and soma, which produces a fairly large postsynaptic conductance increase, gives an effective shunting of the excitatory potentials and action potentials. The synaptic circuitry for such processes includes the basket cells and stratum oriens interneurons which innervate the soma and initial segment of each pyramidal cell[48, 4, 50].

In the KA model of epilepsy, as well as many other chronic and acute models we have found that there is a loss of the early $GABA_A - Cl^-$ mediated IPSP and a failure of inhibitory synaptic function[54, 5]. Iontophoretic application of GABA to the soma and dendrites of CA 1 pyramidal cells in KA lesioned hippocampus indicated that there had been no change in the efficacy of postsynaptic GABA receptors on these cells[6]. This also rules out a possible involvement of the mechanisms described by Inoue et al.[42] where intracellular Ca^{++} ions decrease the affinity of the GABA receptor. Other evidence for a presynaptic mechanism has been presented by Fisher and Alger[25] who observed a reduction in inhibition following the acute application of KA to hippocampal slices. Possible presynaptic mechanisms included the loss of GABA containing interneurones, morphological changes at the inhibitory synapses or depression of the synaptic release of GABA.

Discussion

The purpose of this paper was to introduce some of the membrane and synaptic mechanisms that may be involved in the expression of epileptiform activity in brain tissue. Although most of the experiments have been performed on rats or guinea pigs, very similar electrophysiological data has recently been obtained from in *vitro* studies of the human neocortex[9, 34]. Such experiments performed on neocortical tissue removed from patients with temporal lobe epilepsy will not only allow us to explore the mechanisms underlying the frequently associated drug resistance but will also help us considerably in the evaluation of animal models.

The heterogeneous properties of ionic channels that normally control the activity of an individual cell offer a wide range of potential mechanisms for the genesis of epileptiform activity. However it is this heterogeneity and the application of molecular biology that may enable us to find unique solutions to the treatment of epileptiform activity in different brain regions.

Although much of the classical pharmacology of epilepsy has revolved around the GABA-receptor and its shunting role in synaptic inhibition, the anticonvulsant action of NMDA-receptor antagonists was an important development[22]. The role of the NMDA-receptor and its channel in epileptiform activity is particularly obvious in the hippocampus since it does not appear to contribute to normal excitatory potentials in this region[49, 20, 6, 7]. The membrane and molecular mechanisms responsible for the transition from non-NMDA to NMDA-receptor mediated synaptic potentials may hold the key to other novel treatments of temporal lobe epilepsy[84].

Acknowledgements

The research carried out in my laboratory was generously funded The Wellcome Trust and the MRC. HVW is presently supported under the Wellcome Senior Lectureship Scheme.

References

1. Alger BA (1984) Characteristics of a slow hyperpolarizing synaptic potential in rat hippocampal pyramidal cells in vitro. J Neurophysiol 52: 892–910
2. Alger BA, Williamson A (1988) A transient calcium-dependent potassium component of the epileptiform burst after-hyperpolarization in rat hippocampus. J Physiol 399: 191–205
3. Andrade R, Nicoll RA (1987) Pharmacologically distinct actions of serotonin on single pyramidal neurones of the rat hippocampus recorded in vitro. J Physiol 394: 99–124
4. Ashwood TJ, Lancaster B, Wheal HV (1984) In vivo and in vitro studies on putative interneurons in the rat hippocampus: Possible mediators of feed-forward inhibition. Brain Res 293: 279–291
5. Ashwood TJ, Lancaster B, Wheal HV (1986) Intracellular electrophysiology of CA 1 pyramidal neurones in slices of the kainic acid lesioned hippocampus. Exp Brain Res 62: 189–198
6. Ashwood TJ, Wheal HV (1986) Extracellular studies on the role of N-methyl-D-asparate receptors in epileptiform activity recorded from the kainic acid-lesioned hippocampus. Neurosci Letts 67: 147–152
7. Ashwood TJ (1987) The expression of N-methyl-D-asparate-receptor-mediated component during epileptiform synaptic activity in the hippocampus. Br J Pharmacol 91: 815–822
8. Avoli M, Louvel J, Pumain R, Oliver R (1987) Seizure like discharges induced by lowering [Mg^{++}]o in the human epileptogenic neocortex maintained in vitro. Brain Res 417: 199–203
9. Avoli M, Oliver A (1987) Bursting in human epileptogenic neocortex is depressed by an N-methyl-D-asparate antagonist. Neurosci Lett 76: 249–254
10. Avoli M, Oliver A (1989) Electrophysiological properties and synaptic responses in the deep layers of the human epileptogenic neocortex in vitro. J Neurophysiol 61: 589–606
11. Benardo LS, Masukawa LM, Prince DA (1982) Electrophysiology of isolated hippocampal pyramidal dendrites. J Neurosci 2: 1614–1622
12. Benardo LS, Prince DA (1982) Cholinergic excitation of mammalian hippocampal pyramidal cells. Brain Res 249: 315–331
13. Bowery NG (1982) Balcofen: 10 years on. TIPS 3: 400–403
14. Brown DA, Adams PR (1980) Muscarinic suppression of a novel voltage-sensitive K$^+$-current in a vertebrate neurone. Nature 283: 673–676
15. Brown DA, Griffith WH (1983) Persistent slow inward calcium current in voltage-clamped hippocampal neurones of the guinea-pig. J Physiol 337: 303–320
16. Cole AE, Nicoll RA (1983) Acetylcholine mediates a slow synaptic potential in hippocampal pyramidal cells. Science 221: 1299–1301
17. Cole AE, Nicoll RA (1984 a) Characterization of a slow cholinergic postsynaptic potential recorded in vitro from rat hippocampal pyramidal cells. J Physiol 352: 173–188
18. Cole AE, Nicoll RA (1984 b) The pharmacology of cholinergic excitatory responses in hippocampal pyramidal cells. Brain Res 305: 283–290
19. Colino A, Halliwell JV (1987) Differential modulation of three separate K-conductances in hippocampal CA 1 neurons by serotonin. Nature 328: 73–77
20. Collingridge GL, Kehl SJ, McLennan H (1983) Excitatory amino acids in synaptic transmission in the Schaffer collateral-commissural pathway of the rat hippocampus. J Physiol 334: 33–46
21. Connor JA, Wadman WJ, Hockberger PE, Wong RKS (1988) Sustained dendritic gradients of Ca^{++} induced by excitatory amino acids in CA 1 hippocampal neurons. Science 240: 649–653
22. Croucher MJ, Collins JF, Meldrum BS (1982) Anticonvulsant action of excitatory amino acid antagonists. Science 216: 899–901
23. Dingeldine R (ed) (1986) Brain slices. Plenum Press, New York, 440p

24. Dutar P, Nicoll RA (1988) A physiological role for GABA B receptors in the central nervous system. Nature 332: 156–158

25. Fisher RS, Alger BF (1984) Electrophysiological mechanisms of kainic acid induced epileptiform activity in the rat hippocampal slice. J Neurosci 4: 1312–1323

26. Fox AP, Nowycky MC, Tsien RW (1987) Single-channel recordings of three types of calcium channels in chick sensory neurones. J Physiol 394: 173–200

27. Gahwiler BH, Brown DA (1985) GABA B-receptor-activated K^+ current in voltage clamp CA 3 pyramidal cells in hippocampal cultures. Proc Natl Acad Sci USA 82: 1558–1562

28. Gustafsson B, Galvan M, Grafe P, Wigstrom H (1982) A transient outward current in a mammalian central neurone blocked by 4-aminopyridine. Nature 299: 252–254

29. Gustafsson B, Wigstrom H (1981) Evidence for two types of afterhyperpolarization in CA 1 pyramidal cells in the hippocampus. Brain Res 206: 462–468

30. Haas HL, Green RW (1984) Adenosine enhances afterhyperpolarization and accommodation in hippocampal pyramidal cells. Pflügers Archiv 402: 244–247

31. Haas HL, Green RW (1986) Effects of histamine on hippocampal pyramidal cells of the rat in vivo. Exp Brain Res 62: 123–130

32. Haas HL, Konnerth A (1983) Histamine and noradrenaline decrease calcium-activated potassium conductance in hippocampal pyramidal cells. Nature 302: 432–434

33. Halliwell JV (1983) Caesium-loading reveals two distinct Ca-currents in voltage-clamped guinea-pig hippocampal neurone in vitro. J Physiol 341: 10–11 P.

34. Halliwell JV (1989) Cholinergic responses in human neocortical neurones. In: Frotscher M, Misgeld U (eds) Central cholinergic transmission. Birkhauser Verlag AG, Therwil, (in press)

35. Halliwell JV, Adams PR (1982) Voltage-clamp analysis of muscarinic excitation in hippocampal neurones. Brain Res 250: 71–92

36. Halliwell JV, Othman LB, Pelchen-Matthews A, Dolly JO (1986) Central action of dendrotoxin selective reduction of a transient K-conductance in hippocampus and binding to localized acceptors. Proc Natl Acad Sci USA 83: 493–497

37. Heinemann U, Konnerth A, Pumain R, Wadman WJ (1986) Extracellular calcium and potassium concentration changes in chronic epileptic brain tissue. In: Delgado-Escueta, Ward, Woodbury, Porter (eds) Advances in neurology 44. Raven Press, New York, pp 641–661

38. Herron CE, Williams R, Collingridge GL (1985) A selective N-methyl-D-asparate antagonist depresses epileptiform activity in rat hippocampal slices. Neurosci Lett 61: 225–260

39. Hille B (1984) Ionic channels of excitable membranes. Sinauer Associates Inc. U.S.A

40. Hodgkin AL, Huxley AF (1952) The components of membrane conductance in the giant axon of Loligo. J Physiol 117: 500–544

41. Hotson JR, Prince DA (1980) A Ca-activated hyperpolarization follows repetitive firing in hippocampal neurons. Brain Res 250: 71–92

42. Inoue M, Oomura Y, Yakushiju T, Akaike N (1986) Intracellular calcium ions decrease the affinity of the GABA receptor. Nature 324: 156–158

43. Johnston D, Hablitz JJ, Wilson (1980) Voltage clamp discloses slow inward current in hippocampal burst-firing neurone. Nature 286: 391–393

44. Jones RSG, Heinemann U (1987) Abolition of the orthodromically evoked IPSP of CA 1 pyramidal cells before the EPSP during washout of calcium from hippocampal slices. Exp Brain Res 65: 676–680

45. Kawasaki K, Czeh G, Somjen GG (1988) Prolonged exposure to high potassium concentration results in irreversible loss of synaptic transmission in hippocampal tissue slices. Brain Res 457: 322–329

46. Kerkut GA, Wheal HV (eds) (1984) Electrophysiology of isolated mammalian CNS preparations. Academic Press, London New York, 402 p

47. Kerr DIB, Ong J, Prager RH, Gynther BD, Curtis DR (1987) Phaclofen: a peripheral and central baclofen antagonist. Brain Res 405: 150–154

48. Knowles D, Schwartzkroin PA (1981) Local circuit synaptic interactions in hippocampal brain slices. J Neurosci 1: 318–322

49. Koerner JF, Cotman CW (1982) Response of Schaffer collateral CA 1 pyramidal cell synapses of the hippocampus to analogues of acidic amino acids. Brain Res 251: 105–115

50. Lacaille J-C, Mueller AL, Kunkel DD, Schwartzkroin PA (1987) Local circuit interactions between oriens/alveus interneurons and CA 1 pyramidal cells in hippocampal slices: Electrophysiology and morphology. J Neurosci 7 (7): 1979–1993

51. Lancaster B, Adams PR (1986) Calcium-dependent current generating the afterhyperpolarization of hippocampal neurons. J Neurophysiol 55: 1286–1292

52. Lancaster B, Nicoll RA (1987) Properties of two calcium-activated hyperpolarizations in rat hippocampal neurones. J Physiol 389: 187–204

53. Lancaster B, Wheal HV (1982) A comparative histological and electrophysiological study of some neurotoxins in the rat hippocampus. J Comp Neurol 211: 105–114

54. Lancaster B, Wheal HV (1984) The synaptically evoked late hyperpolarization in hippocampal CA 1 pyramidal cells is resistant to intracellular EGTA. Neurosci 12: 267—75

55. Llinas R (1988) The intrinsic electrophysiological properties of mammlian neurons: insights into central nervous system function. Science 242: 1654–1663

56. Llinas R, Sugimori M (1980) Electrophysiological properties of in vitro Purkinje cell somata in mammalian cerebellar slices. J Physiol 305: 171–195

57. MacVicar BA (1985) Depolarizing prepotentials are Na^+ dependent in CA 1 pyramidal neurons. Brain Res 333: 378

58. Madison DV, Malenka RC, Nicoll RA (1986) Phorbol esters block a voltage-sensitive chloride current in hippocampal pyramidal cells. Nature 321: 695–697

59. Madison DV, Nicoll RA (1982) Noradrenaline blocks accommodation of pyramidal cell discharge in the hippocampus. Nature 299: 636–638

60. Madison DV, Nicoll RA (1984) Control of repetitive discharge of rat pyramidal neurones in vitro. J Physiol 354: 319–331

61. Malenka RC, Nicoll RA (1986) Dopamine decreases the calcium-activated afterhyperpolarization in hippocampal CA 1 pyramidal cells. Brain Res 379: 210–215

62. Margerison JM, Corsellis JAN (1966) Epilepsy and the temporal lobes. A clinical, electroencephalographic and neuropathological study of the brain in epilepsy with particular reference to the temporal lobes. Brain 89: 499–530

63. Meldrum BS, Croucher MJ, Badman G, Collins JF (1983) Antiepileptic action of excitatory amino acid antgonists in the photosensitive baboon *Papio papio*. Neurosci Lett 39: 101–104

64. Miles R, Wong RKS (1984) Unitary inhibitory synaptic potentials in the guinea pig hippocampus in vitro. J Physiol 356: 97–113

65. Nadler JV, Perry BW, Cotman C (1978) Intraventricular kainic acid preferentially destroys hippocampal pyramidal cells. Brain Res 205: 676–677

66. Newberry NR, Nicoll RA (1984) A bicuculline-resistant inhibitory postsynaptic potential in rat hippocampal pyramidal cells in vitro. J Physiol 348: 239–254

67. Nowak L, Bregestovski P, Ascher P, Herbet A, Prochiantz A (1984) Magnesium gates glutamate-activated channels in mouse central neurons. Nature 307: 462–465

68. Pellmar TC (1986) Histamine decreases calcium-mediated potassium current in guinea-pig hippocampal CA 1 pyramidal cells. J Neurophysiol 55: 727–738

69. Rudy B (1988) Diversity and ubiquity of K channels. Neuroscience 25: 729–749

70. Rutecki PA, Lebeda FJ, Johnston D (1985) Epileptiform activity induced by changes in extracellular potassium in hippocampus. J Neurophysiol 54: 1363–1374

71. Schwartzkroin PA, Slawsky M (1977) Probable calcium spikes in hippocampal neurons. Brain Res 135: 157–161

72. Schwartzkroin PA, Stafstrom CE (1980) Effects of EGTA on the calcium-activated afterhyperpolarization in hippocampal CA 3 pyramidal cells. Science 210: 1125–1126

73. Schwartzkroin PA, Wheal HV (eds) (1986) Electrophysiology of epilepsy. Academic Press, London New York, 420 p

74. Segal M, Barker JL (1984) Rat hippocampal neurones in culture: potassium conductances. J Neurophysiol 51: 1409–1433

75. Soltesz I, Haby M, Leresche N, Crunelli V (1988) The GABA B antagonist phaclofen inhibits the late K$^+$ dependent IPSP in cat and rat thalamic and hippocampal neurones. Brain Res 448: 351–354

76. Somjen GG, Aitken PK, Giacchino JL, McNamara JO (1986) Interstitial ion concentrations and paroxysmal discharges in hippocampal formation and spinal cord. In: Delgado-Escueta, Ward, Woodbury, Porter (eds) Advances in neurology 44, Raven Press, New York, pp 663–680

77. Stafstrom CE, Schwindt MC, Chubb WE, Crill WE (1985) Properties of persistent sodium conductance and calcium conductance of layer V neurons from cat sensorimotor contex in vitro. J Neurophysiol 53: 153–170

78. Storm JF (1986) Evidence that C-current and A-current contribute to repolarization of the action potential in CA 1 pyramidal cells of rat hippocampus. Neurosci Abstracts 12: 794

79. Storm JF (1987) Action potential repolarization and a fast afterhyperpolarisation in rat hippocampal pyramidal cells. J Physiol 395: 733–759

80. Swann JW, Smith KL, Brady RJ (1986) Extracellular K$^+$ accumulation during penicillin-induced epileptogenesis in the CA 3 region of immature rat hippocampus. Dev Brain Res 30: 243–255

81. Traynelis SF, Dingledine R (1988) Potassium-induced spontaneous electrographic seizures in the rat hippocampal slice. J Neurophysiol 59: 259–276

82. Tsien RY (1980) New calcium indicators and buffers with high selectivity against magnesium and protons: design, synthesis and properties of prototype structures. Biochem 19: 2396–2404

83. Turner DA (1985) The influence of feed-forward IPSPs on small EPSPs in hippocampal DA 1 pyramidal cells in vitro: timing and interactions. Neurosci Lett 22: S 511

84. Turner DA, Wheal HV (1988) Components of subthreshold synaptic potentials in CA 1 pyramidal neurones from kainic acid lesioned rat hippocampus in vitro. J Physiol 400: 50 P

85. Wadman WJ, Heinemann U (1983) Laminar profiles of changes in extracellular potassium and calcium concentration in slices obtained from kindled rats. In: Baldy-Mouliner, Ingvar, Meldrum (eds) Cerebral blood flow, metabolism and epilepsy. John Libbey, London, pp 315–323

86. Watkins JC, Evans RH (1981) Excitatory amino acid transmitters. Annu Rev Pharmacol Toxicol 21: 165–204

87. Wheal HV (1989) Function of synapses in the CA 1 region of the hippocampus: their contribution to the generation or control of epileptiform activity. Comp Biochem Physiol 93 A: 211–220

88. Yarri Y, Konnerth A, Heinemann U (1983) Spontaneous epileptiform activity of CA 1 hippocampal neurons in low extracellular calcium solutions. Exp Brain Res 51: 153–156

Correspondence: H. V. Wheal, M.D., Department of Neurophysiology, University of Southampton, Bassett Crescent East, Southampton, S09 3TU, U.K.

Acta Neurochirurgica, Suppl. 50, 14–18 (1990)

Pathophysiology of Acute Brain Damage Following Epilepsy*

A. Baethmann

Institute for Surgical Research, Ludwig-Maximilians-University of Munich, Federal Republic of Germany

Summary

The possible pathophysiological mechanisms, both intrinsic and systemic, leading to acute brain damage following epilepsy are reviewed. In particular involvement of changes in blood brain barrier, alterations of acid base regulation in the brain, release of a variety of mediator compounds, such as arachidonic acid and glutamate, intracellular influx of calcium ions, and the inhibition of protein synthesis are discussed. Finally, pathophysiology of brain damage following epilepsy is compared with that following ischaemia and hypoglycaemia.

Keywords: Epilepsy; brain damage; pathophysiology.

Overview

Major findings characterizing the pathophysiology of epileptic brain damage are summarized in Table 1. Provided general complications involving the circulatory and ventilatory system do not occur, there is little evidence for the development of energy failure in epilepsy as a mechanism of brain damage in spite of the hypermetabolic state during convulsions. The manifestations of hypermetabolism are not uniform but appear to affect particularly active brain tissue areas, such as the hippocampus and layers of the cerebral cortex. In spite of the maintenance of energy metabolism, marked alterations of brain tissue homeostasis may occur, as for example an intra- and extracellular acidosis from accumulation of lactic acid and release of mediator compounds, such as arachidonic acid. It can be assumed that these abnormalities are particularly severe in selectively vulnerable areas. This is also true for manifestations of functional and structural damage including the loss of nerve cells. Hallmarks are severe swelling of dendrites, intracellular deposition of Ca^{++}, inhibition of protein synthesis, and chronically the development of abnormal dendritic structure, loss of den-

Table 1. *Summary of Major Pathophysiological Findings Concerning Brain-Damage from Epilepsy*

1. Little evidence of involvement of energy failure in spite of hypermetabolic state during ictus
2. However, there is development of brain tissue acidosis, accumulation of lactate, and release of arachidonic acid
3. Manifestations of structural damage and loss of nerve cells are focussed in selectively vulnerable areas, *e.g.* in hippocampus, substantia nigra, and cortex
   - Swelling of dendrites
   - Intracellular deposition of Ca^{++}-ions
   - Inhibition of protein synthesis
   - *Later:* Loss of dendritic spines, deformation of dendrites, microaneurysms of capillaries, *etc.*
4. Opening of blood-brain barrier—*without major oedema formation*—may result from systemic hypertension. The effect is directed at selective areas and is dependent on the seizure inducing agent

dritic spines, and abnormalities of vessels of the microcirculation[14].

Opening of the blood-brain barrier during seizures does not result from mechanisms intrinsic to convulsing brain tissue, nor from alterations of glial elements, but rather from the abrupt, ictal increase of the systemic blood pressure. Yet, edema formation is not a prominent feature in acute epileptic seizures unless widespread structural damage has been induced, as in maintained status epilepticus. Chronic seizures with frequent disruptions of the blood-brain barrier eventually cause permanent functional and structural alterations of the elements of the blood-brain barrier. Then, thickening of the basal membrane of cerebral capillaries is observed as well as an impairment of carrier-dependent substrate transport. Development of selective nerve cell necrosis and, occasionally, of regional infarcts, might

* Supported by Deutsche Forschungsgemeinschaft: Ba 452/6–7.

be explained on the basis of an uncontrollable excitation and intracellular influx of Ca^{++} in association with excessive accumulation of mediator compounds, such as arachidonic acid and glutamate. Under these conditions certain brain tissue areas may incur a failure of the oxygen- and substrate supply and become susceptible to the formation of edema.

Introduction

Development of acute brain damage from seizures is difficult to understand. Both CNS-intrinsic mechanisms, associated with the hyperactivity and hyperexcitability of nerve cells, as well as systemic factors, resulting from the pathophysiologic alterations of the circulatory and ventilatory system, may play a role. A general assumption has been that hyperactive, seizing nerve cells eventually suffer from an impairment of the blood- and O_2-supply leading to energy failure. Recent evidence, however, does not support a major breakdown of the cellular energy supply, at least if the event is limited. Therefore, the hypothesis of a central pathophysiological role of a supply/-demand dysequilibrium cannot be upheld. Various findings demonstrate that actively seizing brain tissue does not suffer from hypoxia nor a sizeable decrease of the energy charge potential. To the contrary, convulsing brain tissue is hyperperfused and hyperoxygenated[9, 15, 17]. This is closely associated with an enhanced glucose consumption as a global measure of the cerebral metabolic and functional activity[1, 8, 18].

The acute PO_2-response and redox state obtained by cytochrome aa_3 measurements at the brain surface of rats during induction of a seizure by i.v. injection of 0.3 mg/kg bicuculline are shown in Fig. 1. The electrical hyperactivity of the brain was associated with an immediate increase of the cerebral PO_2 and oxidation of cytochrome aa_3.

In-vivo measurements of regional metabolic activity have been employed to analyze the functional abnormalities on a topographical basis during epilepsy and in the interictal phase. Between seizures, marked hypometabolism may prevail in patients with partial epilepsy[7]. Focal areas of hypometabolism in patients with complex partial epilepsy, which are considered amenable for surgery might indicate the epileptogenic focus. Removal of such tissue might be beneficial.

Experimental research has provided a wealth of information on acute and chronic pathophysiological alterations in the brain during and after epileptic seizures. Areas of interest among others are the involvement of

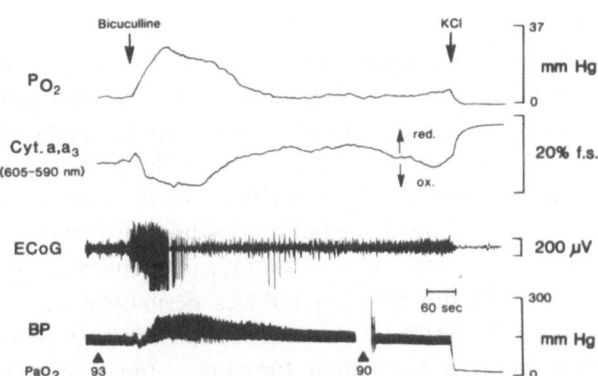

Fig. 1. Increase in cerebral O_2-tension (PO_2) and of cytochrome aa_3 oxidation following induction of a seizure by 0.3 mg/kg bicuculline in a rat. Note the corresponding EEG response (ECoG) as well as the increase in systemic blood pressure. The pulmonary oxygenation was not impaired as demonstrated by a normal arterial PO_2 (from: Kreismann NR *et al* 1987)[10]

blood-brain barrier function, alterations of acid-base regulation in the brain, release of mediator compounds, such as arachidonic acid or glutamate, intracellular influx of Ca^{++}-ions, and global measures such as protein synthesis.

Blood-Brain Barrier Function

Opening of the blood-brain barrier from epileptic seizures induced by administration of electroshocks or other methods appears to be dose-dependent. In a study of Petito (1976)[16] it was found that the number of electroshocks administered to an experimental animal correlated with the extent of barrier damage. Ultrastructural investigations performed in parallel have demonstrated penetration of intravenously administered horseradish peroxidase by apparent enhancement of pinocytosis through the cerebrovascular endothelium[16]. In these studies it was clearly demonstrated that opening of the blood-brain barrier cannot be attributed to the hyperactive brain tissue itself, but rather to the response of the general circulation, particularly to the abrupt increase of the systemic blood pressure. This conclusion was supported by findings of experiments on combination of seizure induction with cordotomy to prevent the systemic blood pressure rise. Opening of the barrier did not occur under these conditions[16].

Similar observations were reported by Nitsch and Klatzo (1983)[13] using different seizure-inducing agents, such as pentylenetetrazol or bicuculline. The authors confirmed that an increase of the systemic blood pressure is required for opening of the blood-brain barrier.

Distinct patterns of barrier opening were found, which appeared to specifically correlate with the agent used for seizure induction. The authors made it clear that on the one hand a systemic blood pressure response was always necessary for opening of the blood-brain barrier, whereas on the other hand its disruption did not always follow if the blood pressure was increased. Long-term studies indicate that chronic epilepsy with frequent seizure episodes induces permanent alterations of the vessels of the microcirculation and of the blood-brain barrier on a functional and structural basis[5, 14]. Thickening of the basal membrane is reported in cerebral capillaries with an increased number of mitochondria in the cerebrovascular endothelium and of interendothelial tight junctions. Hence, substrate transport of, e.g. amino acids or glucose, through the barrier might be limited during convulsions.

Acid-Base State

Measurements of the extracellular pH in brain tissue of experimental animals subjected to seizures have demonstrated a biphasic course of tissue acidosis[19]. Administration of fluorothyl used as seizure-inducing agent led to an immediate decrease of the extracellular pH to values of 7.0 or 6.6. This was followed by brief recovery, however, not to normal but still to an acidotic pH. Extracellular acidosis may prevail for quite a while after termination of seizures. Based on measurements of the extracellular pH together with tissue pCO_2 and other variables, the intracellular pH was calculated. An intracellular acidosis was confirmed which, however, was found to disappear more rapidly again after termination of the seizure, and to be followed by the development of an intracellular alkalosis[19]. Generation of extra- and intracellular acidosis during epilepsy in spite of a practically normal oxidative energy state and unchanged energy charge potential is attributed to the enhancement of aerobic glycolysis leading to the accumulation of lactic acid[19], raising questions on the underlying mechansims. Siesjö and Wieloch (1986)[18] come to the conclusion that subtle and regional perturbations of the energy metabolism in convulsing brain tissue cannot be completely excluded. Thus, cerebral structures that are particularly active during seizures such as the hippocampus, or layers of cerebral cortex, may develop some energy failure triggering further adverse reactions. Release of mediator compounds and an intracellular accumulation of Ca^{++} appear particularly important[2, 12], which altogether may be ultimately responsible for the extinction of selectively vulnerable neurons.

Release of Mediator Compounds

A mechanism of acute brain damage from epilepsy relates to the liberation of noxious mediator compounds in the brain, such as glutamate or arachidonic acid. Activation of lipolysis causing accumulation of arachidonic acid has been reported[2, 18], whereas data on an extracellular accumulation of glutamate are not unequivocal[11]. Nevertheless, protection of nerve cells against epileptic damage by glutamate receptor antagonists has been demonstrated[4]. An important cellular messenger system that is activated by release of mediators such as glutamate or others, is the phosphatidylinositol system. During convulsions receptor-ligand interaction with activation of phospholipase C (PLC) seems to play a role in brain tissue[2]. This process associated with formation of diacylglycerol (DG) and inositol-1, 4, 5,-trisphosphate (IP_3) might be considered as a major step for the release of arachidonic acid after onset of a seizure. The release of arachidonic acid (C20:4) is particularly pronounced during the early phase of epileptic seizures[18]. Activation of the phosphatidylinositol system causing hydrolysis of membrane bound phospholipids might be of major pathophysiological significance as a central mechanism ultimately causing death of selectively vulnerable nerve cells. Cell death might relate to the uncontrolled formation of IP_3 leading to an intracellular release and accumulation of Ca^{++}-ions with its potentially fatal consequences, such as cellular stimulation of lipolysis and proteolysis[12]. Another aspect of cell damage might be associated with the release of diacylglycerol and subsequent activation of protein kinase C, which amongst other effects induces the Na^+/H^+-antiporter, thereby facilitating cell swelling. Marked swelling of nerve cell fibers, for example dendrites in association with an abnormal deposition of Ca^{++}, has been reported in selectively vulnerable structures[12].

It is not clear whether and how inhibition of protein synthesis in epileptic brain tissue in selectively vulnerable areas relates to (a) the accumulation of mediator compounds, (b) the increase of intracellular Ca^{++}, and (c) cell death. Inhibition of protein synthesis from epileptic seizures occurs together with disaggregation of polyribosomes[6]. It was found that the extent of inhibition of protein synthesis, or of polyribosome disaggregation as well as its reversibility correlated with both duration and number of seizures[6]. It may be noted that inhibition of protein synthesis does not follow a uniform pattern involving all brain structures but rather displays specific regional differences. In the cerebral

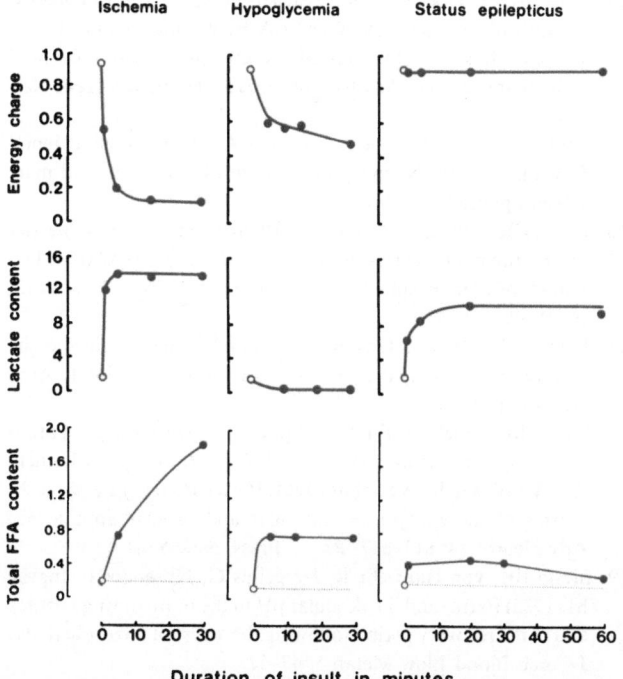

Fig. 2. Comparison of cerebral energy potential and tissue concentrations of lactate and of free fatty acids in (a) complete cerebral ischemia, (b) hypoglycemia, and (c) status epilepticus. The duration of a pathological condition is given on the abscissa. Obviously, complete cerebral ischemia causes the most pronounced alterations, whereas relatively minor changes were elicited by status epilepticus, even after 60 min. Due to the decrease of blood glucose levels, hypoglycemia cannot increase tissue levels of lactic acid. Nevertheless, development of an energy failure and enhancement of lipolysis can be concluded from the decrease of the energy charge potential and increase of free fatty acids in the brain. Status epilepticus is associated with an accumulation of lactic acid in brain tissue in spite of a practically normal energy state. Thus, activation of anaerobic glycolysis might occur independently from an impairment of the oxidative metabolism under these conditions (from: Siesjö BK, Wieloch T, 1986)

cortex and hippocampus, regions of severe inhibition of protein synthesis may alternate with regions of relative maintenance[6].

Comparison of Ischemia, Hypoglycemia and Epilepsy

Auer and Siesjö[1] (1988) have made an interesting approach to compare semiquantitatively the damaging potential of global cerebral ischemia, hypoglycemia, and status epilepticus (cf. Fig. 2). Obviously, global ischemia causes the densest form of brain tissue damage as seen from the complete breakdown of energy metabolism and the most severe accumulation of lactic acid. The degree of free fatty acid accumulation in ischemic brain tissue is another indicator of the severity of brain tissue damage from global ischemia. Hypo-

glycemia is less severe as far as the development of energy failure and release of free fatty acids is concerned, whereas, for obvious reasons, lactic acid is not formed. In status epilepticus, even of 60 min duration, there is virtually no decrease of the energy state indicative of absence of energy failure in spite of an increase of lactic acid formation and release of free fatty acids[18]. Analysis of the extinction of nerve cells in selectively vulnerable structures indicates that ischemia, hypoglycemia, and status epilepticus are all causing a demise of neurons, however, in a clearly distinct manner. Status epilepticus causes destruction of nerve cells in layers 3 and 4 in cerebral cortex, with none in caudate nucleus, or the CA 3 layer of hippocampus. On the other hand, global ischemia destroys selectively vulnerable neurons in the caudate nucleus, CA 3 and subiculum contrary to status epilepticus[18].

Thus, it is a reasonable assumption that there are similarities as well as dissimilarities in the generation of irreversible nerve cell damage by the above conditions. In an attempt to compare the damaging potential of ischemia, hypoglycemia, and epilepsy on a quantitative level the authors suggest that an equivalent degree of damage, as quantified by the extent of nerve cell necrosis, is produced by (a) 2–10 min global cerebral ischemia, (b) 10–20 min hypoglycemia, and (c) 45–120 min status epilepticus[1, 18].

Acknowledgements

The secretarial and technical assistance of Isolde Juna and Helga Kleylein is gratefully appreciated.

References

1. Auer RN, Siesjö BK (1988) Biological differences between ischemia, hypoglycemia, and epilepsy. Ann Neurol 24: 699–707
2. Bazan NG, Rakowski H (1970) Increased levels of brain free fatty acids and electroconvulsive shock. Life Sci 9: 501–507
3. Bazan NG, Birkle DL, Tang W, Reddy TS (1986) The accumulation of free arachidonic acid, diacylglycerols, prostaglandins, and lipoxygenase reaction products in the brain during experimental epilepsy. In: Delgado-Escueta AV, Ward AA Jr, Woodbury DM, Porter RJ (eds) Basic mechanisms of the epilepsies – molecular and cellular approaches. Adv Neurol 44: 879–902. Raven Press, New York
4. Chapman AG, Engelsen B, Meldrum BS (1987) 2-amino-7-phosphonoheptanoic acid inhibits insulin-induced convulsions and striatal aspartate accumulation in rats with frontal cortical ablation. J Neurochem 49: 121–127
5. Cornford EM, Oldendorf W (1986) Epilepsy and the blood-brain barrier. In: Delgado-Escueta AV, Ward AA Jr, Woodbury DM, Porter RJ (eds) Basic mechanisms of the epilepsies – molecular and cellular approaches. Adv Neurol 44: 787–812. Raven Press, New York

6. Dwyer BE, Wasterlain CG, Fujikawa DG, Yamada L (1986) Brain protein metabolism in epilepsy. In: Delgado-Escueta AV, Ward AA Jr, Woodbury DM, Porter RJ (eds) Basic mechanism of the epilepsies – molecular and cellular approaches. Adv Neurol 44: 903–918. Raven Press, New York

7. Engel J Jr, Kuhl DE, Phelps ME, Rausch R, Nuwer M (1983) Local cerebral metabolism during partial seizures. Neurology 33: 400–413

8. Evans MC, Meldrum BS (1984) Regional brain glucose metabolism in chemically-induced seizures in the rat. Brain Res: 297: 235–245

9. Jöbsis FF, O'Connor M, Vitale A, Vreman H (1971) Intracellular redox changes in functioning cerebral cortex. I. Metabolic effects of epileptiform activity. J Neurophysiol 34: 735–749

10. Kreisman NR, Hodin RA, Brizzee BL, Rosenthal M, Sick TJ, Busto R, Ginsberg MD (1987) Seizure-associated pulmonary edema and cerebral oxygenation in the rat. J Appl Physiol 62 (2): 658–667

11. Lehmann A, Hagberg H, Jacobson I, Hamberger A (1985) Effects of status epilepticus on extracellular amino acids in the hippocampus. Brain Res 359: 147–151

12. Meldrum BS (1986) Cell damage in epilepsy and the role of calcium in cytotoxicity. In: Delgado-Escueta AV, Ward AA Jr, Woodbury DM, Porter RJ (eds) Basic mechanisms of the epilepsies – molecular and cellular approaches. Adv Neurol 44: 849–855. Raven Press, New York

13. Nitsch C, Klatzo I (1983) Regional patterns of blood-brain barrier breakdown during epileptiform seizures induced by various convulsive agents. J Neurol Sci 59: 305–322

14. Paul LA, Scheibel AB (1986) Structural substrates of epilepsy. In: Delgado-Escueta AV, Ward AA Jr, Woodbury DM, Porter RJ (eds) Basic mechanisms of the epilepsies – molecular and cellular approaches. Adv Neurol 44: 775–786. Raven Press, New York

15. Penfield W, von Santha K, Cipriani A (1939) Cerebral blood flow during induced epileptiform seizures in animals and man. J Neurophysiol 2: 257–267

16. Petito CK, Schaefer JA, Plum F (1976) The blood-brain barrier in experimental seizures. In: Pappius HM, Feindel W (eds) Dynamics of brain edema. Springer, Berlin Heidelberg New York, pp 38–49

17. Plum F, Rosner JB, Troy B (1968) Cerebral metabolic and circulatory responses to drug-induced convulsions in animals. Arch Neurol 18: 1–13

18. Siesjö BK, Wieloch T (1986) Epileptic brain damage: Pathophysiology and neurochemical pathology. In: Delgado-Escueta AV, Ward AA Jr, Woodbury DM, Porter RJ (eds) Basic mechanisms of the epilepsies – molecular and cellular approaches. Adv Neurol 44: 813–847. Raven Press, New York

19. Siesjö BK, von Hanwehr R, Nergelius G, Nevander F, Ingvar M (1985) Extra- and intracellular pH in the brain during seizures and in the recovery period following the arrest of seizure activity. J Cereb Blood Flow Metab 5: 47–57

Correspondence: A. Baethmann, M.D., Institute for Surgical Research, Ludwig-Maximilians-University of Munich, Klinikum Großhadern, D-8000 München 70, Federal Republic of Germany.

Acta Neurochirurgica, Suppl. 50, 19 (1990)

Sampling Extracellular Amino Acids from the Brain Surface During Epilepsy Surgery – Abstract of Discussion

A. Hamberger, B. Nyström[1], H. Silfvenius[2], and A. Hedström[3]

Departments of [1] Neurobiology, [2] Neurosurgery, and [3] Clinical Neurophysiology, University of Göteborg, Sweden

The concentration of extracellular amino acids has been monitored from the exposed brain surface during epilepsy surgery with the aim of investigating a possible correlation with the electrocortigraphic (ECoG) activity. Sampling has been performed in 17 patients, 6 children and 11 adults. Epilepsy surgery was carried out on nontumoral (n = 7) or nonexpanding tumoral (n = 19) cases.

Sampling was done by slowly perfusing Krebs-Ringer bicarbonate medium (2.5 ml/min) into 0.3 mm dialysis tubing in order to equilibrate with the extracellular amino acids. The tubing was moulded into a silicon patch between holes for carbon ball electrodes to record the ECoG. The tubing was gently and continously in contact with the pial surface of the brain. The basic method for brain microdialysis was originally described by Ungerstedt and coworkers for experimental studies and intraparenchymal analysis. It has been used for experimental research and developed for clinical use in our laboratory. Routinely, two surface dialytrodes were used, one placed on a "normal" area, the other on an epileptogenic cortical area. Each was held in place with two ECoG electrodes. Sampling was done during 15–20 min which allowed the collection of 3–4 samples. The pial surface was not moistened during the sampling period. The samples were stored at −80 °C until analysed for amino acids with liquid chromatography. The technique was evaluated by comparing the amino acid concentration in plasma, blood and CSF with that on the brain surface. The samples from the brain surface could not represent a mixture of CSF and blood, but characterized the extracellular phase of the pial surface. A considerable difference between intra- and extracellular brain amino acids was obviously at hand. In the latter compartment glutamine, was relatively prominent. Glutamate, phenylalanine, and taurine, were followed over three epochs (5–10–15 min). Glutamate and phenylalanine decreased by some 60%, while taurine remained constant. This was interpreted as reflecting the different release and uptake mechanisms for amino acids. The results from the last sample was routinely used, since the concentration of amino acids had levelled off to a relative steady state at that time.

The mean values for amino acids in normal and epileptic (interictal spiking) cortical areas were compared in five patients. The epileptic regions, mainly lateral temporal cortex-receiving projected abnormality from anterior mesial structures – had higher concentrations of ethanolamine than normal cortical regions. The functional significance of this is still obscure. The concentrations of serine, and glutamine were lower in the epileptic area than in normal cortex. Asparate decreased less during the sampling period (5–20 min) in the epileptic areas than in the normal areas.

Amino acids were also monitored in other patients subjected to neurosurgery: two had gliotic-atrophic epileptic regions, the third a superficial low grade astrocytoma. Taurine was elevated in the first two. The astrocytoma pattern was very similar to that of blood, suggesting damage to the blood-brain barrier. The correlation of the amino acid concentration with interictal ECoG spiking has so far not revealed any dramatic findings. Dialytrodes have been inserted into the CNS parenchyma to collect from the epileptic mesial temporal structures.

Correspondence: A. Hamberger, M.D., Ph.D., Department of Neurobiology, University of Göteborg, Sweden.

Acta Neurochirurgica, Suppl. 50, 20–25 (1990)
© by Springer-Verlag 1990

Neuropathology of Epilepsy

A. Mouritzen Dam

Neurological Research Laboratory, University Hospital Hvidovre Copenhagen, Hvidovre, Denmark

Summary

The neuropathology of symptomatic and idiopathic epilepsy is presented. New methods of investigation have disclosed that more cases of epilepsy are symptomatic – even hippocampal sclerosis is now detectable in vivo by MR-scan. It still remains to be defined whether hippocampal sclerosis is the cause or the effect of epilepsy. The present survey focusses upon the delineation of hippocampal neuron loss with a critical discussion of various hypotheses drawn from various human and experimental studies. In particular the phenomenon of selective and delayed neuron loss are explored with regard to future neurosurgical approaches.

Keywords: Hippocampus; neuron loss, epilepsy; neuropathology.

Introduction

Pathology of the brain is not always accompanied by epileptc seizures. When seizures occur in relation to brain pathology, they may be caused by the pathology but some types of pathology may be caused by the epileptic seizures.

In expanding lesions such as brain tumors and malformations, there seems to be a causal relationship whereby removal of the pathology eliminates the epileptic seizures. This is true only in 50%[18]. The scar itself may continue to cause seizure activity but seizures may also continue due to a mirror focus or because another region in the brain was already the pacemaker of the epileptic acitivty. Any type of cerebral pathology, including shrinkage of the brain due to infarction, trauma or infection may all give rise to epilepsy. The site of the lesion is important. Any structural lesion probably has to involve the cortex[18].

Symptomatic epilepsy is the group where the clinical investigation often discloses focal symptoms and signs and where adequate pathology is discovered. Idiopathic epilepsy is the group of epilepsies often with hereditary disposition, without neurological symptoms and signs or pathology. This division may not be so clearcut. It is difficult to ensure that no pathology is present. Furthermore, improvements in diagnostic methods (video-monitoring, PET-, SPECT- and MR-scans) almost certainly will reduce the number of "idiopathic" epilepsies. Finally it is not yet known whether pathology such as hippocampal sclerosis is the cause or the effect of epilepsy.

Symptomatic Epilepsy

Brain tumors, congenital abnormalities, degenerative processes and vascular accidents are common symptomatic causes of epilepsy. Cerebral atrophy, diagnosed in vivo after the introduction of the CT-scan in particular, may be associated with repeated epileptic seizures. General atrophy is aften diagnosed but the epileptogenic mechanism is probably related to focal lesions. It is difficult to be precise about the incidence of the different types of abnormalities. The figures also vary according to the source and age range of the patients. In 1963, Serafetinides[30] disclosed pathology *in vivo* in 25% of patients with late onset epilepsy. That number was increased to 60% in a similar study[24] in 1985.

Brain tumor is one of the causes most emphasized when fits develop in adults[18]. 70–80% of the patients with tumors will, in particular when the tumor is localized near the central sulcus, develop epileptic fits and the fits will be the presenting sign of the tumors in 75% of the cases. Slowly growing tumors give fits more often than those with rapid, invasive growth. The incidence of seizures is about 60% for oligodendroglioma and astrocytoma, approximately 40% for meningioma and only 30% for glioblastoma[13,14,18]. It is also worth mentioning that the removal of a tumor will abolish the seizures in 50%, but another 10% will develop seizures after operation for brain tumor without having any seizures before the operation[14].

Another relatively common cause of epilepsy starting after the age of 50 years is cerebral infarction. The incidence has probably been underestimated. Previous studies mention only a few percent but incidences up to 10% and more have recently been published[5,31]. Some studies include single seizures which is not epilepsy, and the incidence rates depend on several factors, one being the time of observation. At present an incidence of 7% seems justified[31]. Approximately 60% of these patients will develop seizures within 2 years after the stroke[5] which implies that a longer observation time will raise the possibility of an additional 40%. Epileptic seizures, in relation to strokes as well as tumors, occur mainly when cerebral cortex is involved, and in particular when the territories around the carotid and middle cerebral arteries are involved[18].

Idiopathic Epilepsy

We are left with a substantial group of patients where no lesion in the brain is disclosed in vivo with the methods used at present and where no disturbances are found outside the brain, such as disturbances of body fluids and hypoglycemia, which are known to be able to cause seizures. Postmortem macroscopic investigation may reveal pathology which for one reason or another was not visible in vivo, but the major gains will be obtained by microscopic examination.

The subtle changes in the brains from patients with epilepsy are of particular interest. The answer to the question of whether such changes are the cause or the effect of the epilepsy might clarify the aetiology of such epilepsy. Neuron loss, gliosis and shrinkage of brain tissue are the pathological aspects which are dealt with in the microscopic investigations.

Hippocampal Pathology

The hippocampus, approximately 5 cm long, is enclosed in the temporal lobe, behind the amygdala nucleus and bordering the lateral ventricle on its mesial side. The structure delineates itself from the neocortex, containing pyramidal cells as a convoluted band ending in a comma-shaped configuration of granule cells of the fascia dentata. The structure is proportionately larger in lower animals but very alike in those species involved in epilepsy research.

It is more than 150 years ago since atrophy of the temporal lobes was regularly observed and described in patients who had suffered from epilepsy[2]. Many years passed before further details were obtained by means of the microscope, which was only in common use from the middle of the 19th century. From that time on it was realized that neuron loss was one major feature of the pathology (Fig. 1 A and B). Sommer[32] described circumscribed neuronal loss in one part of the pyramidal cell band in the hippcampus, which he named Ammon's horn sclerosis due to the resemblance between the hippocampal convoluted structure and the Egyptian god Ammon's helmet. The name "Sommer's sector" was applied to the area Sommer described. The delineation and terminology have been little changed since. Field H 1[29] or field Ca 1[28] are some other names given which are used in modern research in this area. Some details about the hippocampal pathology include the unilateral occurrence mentioned by Bouchet and

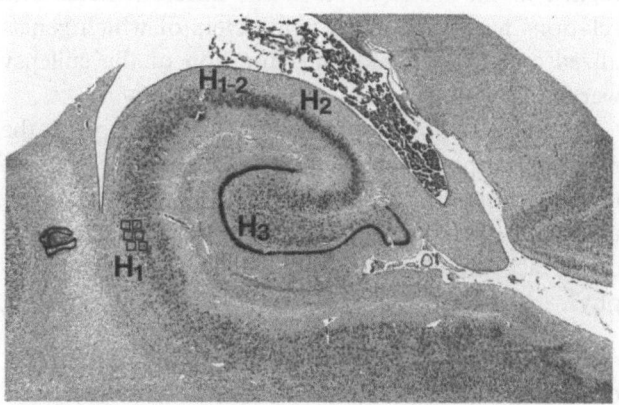

A

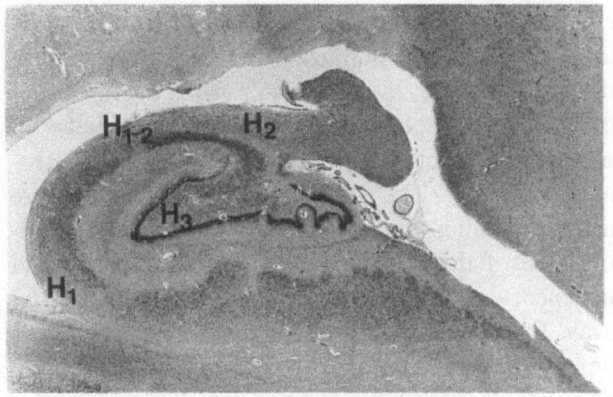

B

Fig. 1. Microscopic version of a cross section of the hippocampus. A) Normal neuronal population. B) Neuron loss apparent in H 1 and H 3 even without quantitation (= hippocampal sclerosis)

Cazauvieilh[2] and later confirmed by Meynert[22] and Bratz[3].

The neurons of field H 2 have received the reputation of being well preserved[3]. The variability in degree of neuron loss throughout the hippocampus was emphasized early[22]. A subdivision of Ammon's horn sclerosis was proposed by Corsellis[19]: firstly, the "classical" type[19,22] with unilateral occurrence of H 1 neuron loss, and secondly, the "end plate" sclerosis or mesial temporal sclerosis with more widespread damage bilaterally, inside as well as outside the hippocampal structure.

The pathology described as Ammon's horn sclerosis was very heterogeneous before this time. Some described severe neuron loss, others only a tendency to decreased neuron density and others mainly gliosis. A further attempt to clarify the pattern of the pathology was introduced by quantitation of the neurons. Quantitation of nerve cells in relation to epilepsy has been applied to cerebellum[7] and the temporal lobe[4,23]. A systemic counting of neurons in the hippocampus was performed in patients with epilepsy. One study was performed postmortem in patients who had suffered from different types of epilepsy. The results were evaluated in relation to different clinical parameters including the different types of epileptic seizures which the patients had suffered from[23].

The various hypotheses regarding the cause of neuron loss and damage in hippocampus included at that time damage of prenatal origin[32], damage during partum or in early life due to hypoxia accompanying various conditions such as status epilepticus or febrile convulsions[8,33]. Another quantitative study[4] was applied to resected temporal lobes from patients who were operated upon due to intractable temporal lobe seizures.

Hughlings Jackson[12] was probably the first clinician who proposed the close relationship between the temporal lobes and epileptic seizures with sensory modalities such as smell, taste and fear. After the invention of the electroencephalograph[10] a new era dawned with regard to pathology also. The operative era was introduced in 1941[26]. The study material was often sparse in the beginning, as the operation was merely a continuation of the resective surgery already performed on neocortex and a major part of the tissue was sucked away. By the 1950s surgeons had developed the technique of surgical "en bloc" resection of the temporal lobe, including the hippocampus, which allows microscopic evaluation of the entire specimen[8].

The quantitative methodology applied was the same in the two studies mentioned. The number of neurons were counted by means of a light microscope at a magnification of approximately × 3–500. A frame was applied on the structure in sections of 10 microns. The frame was moved from side to side and up and down to cover a predefined total area within each pyramidal field as well as granule area. The various pitfalls in studies of brain tissue, such as postmortem decomposition and shrinkage due to the pathological procedures and errors related to the counting methods, were avoided or evaluated as much as possible. The pathology was evaluated in comparison with control cases in order to know the variations in non-epileptic related cases. The pattern of neuron loss which was found in the postmortem study of a non-selected group with different types of epilepsy underlined the importance of quantitation. Neuron loss was most prominent in field H 3, with field H 1 taking second place. There was difference between the two sides but strictly unilateral neuron loss was seen in one case only, and the side-to-side difference was otherwise unsystematic.

Such quantitation disclosed for the first time a difference in neuron density between the anterior and posterior part of the hippocampus. The neuron density was lower at the anterior end in non-epileptic cases also, but more marked in relation to epilepsy. As a whole, a much more varied pattern of the neuron loss was disclosed by the application of the counting procedure. It should be emphasized that the study was not related to temporal lobe epilepsy in particular, and those cases with temporal lobe epilepsy did not have a different pattern of neuron loss in comparison with the others in the group where generalized convulsions were the most common type of epileptic seizures. The application of the counting procedure to the postmortem study supplied further details in particular with regard to the distribution in the different fields and relations to some clinical parameters of which generalized convulsions and long duration of the epilepsy were most important.

The application of the quantitative method to the resected temporal lobes did support the specificity of the neuron loss in field H 1 and, of particular importance, the results were related to the predictive value of the accompanying presurgical stereoencephalography[4].

Cerebellum

The Purkinje cells in the cerebellum exert a role in the pathology of epilepsy. Loss of Purkinje cells was described in patients with epilepsy in relation to multiple generalized convulsion[29, 33]. Quantitation of Purkinje cells in patients with epilepsy was first applied in 1972[7] in a study which compared treatment with diphenylhydantoin, suspected of causing irreversible damage to the Purkinje cells. The Purkinje cells were counted in a light microscope as the number of cells per length (mm). The study supported the relationship already hypothesized between loss of Purkinje cells and many generalized convulsions[7].

Cortex Cerebri

Microdysgenesis was looked upon as a maldevelopment of some undefined importance for epilepsy[1]. Localized cortical dysplasia/dystopia and clustering of nerve cells are found in cerebrum and cerebellum. The first description was related to focal cortical epilepsy and to temporal lobe epilepsy. It was previously described in cortex of temporal lobectomies but also in the cerebellum of brains related to symptomatic epilepsy with hippocampal neuron loss. It has recently been described in relation to idiopathic epilepsies[20]. The significance of this phenomenon in relation to pathological changes in epilepsy is not yet solved.

Experimental Seizures

Epileptic seizures can be provoked in animals. Experimental studies of epilepsy have been concerned with the elementary properties of neurons. The first experimental model of epilepsy were seizures produced by injection of alumina cream into cortex of monkeys[16]. Focal seizures with secondary generalization developed after a certain delay and continued for months and years. Seizures elicited by focal electrical stimulation were explored in another model[9], that showed the difference in epileptogenicity in the various regions of the brain. The particular susceptibility of the hippocampus was evident and so was the low threshold of the sensory-motor cortex. The cerebellum was the only region where seizures could not be elicited. The inhibitory role of cerebellum, in particular of the Purkinje cells, was later confirmed. Electrical stimulation applied to cerebellum as a trial of treatment[6] failed, however. Systemic injection of different convulsants resulted in generalized repeated seizures or status epilepticus[21]. These studies were the first studies to show neuron damage, in particular in the hippocampus, resembling the pathology in patients with epilepsy. A correlation between neuron loss in Sommer's sector and duration of the convulsions was demonstrated. The similarity between damage related to epilepsy and ischemia was underlined[27].

Quantitative morphology has also been applied to experimental epilepsy. Electroconvulsive seizures were elicited in rats in a number so that the total seizure-time went up to 3 hours. Sacrifice was postponed for some weeks. The results of quantitative morphology disclosed no irreversible damage to the neurons in hippocampus nor the Purkinje cells[25], in contrast with the spontaneous seizures in the mongolian gerbils, where neuron loss occurred. The seizures in these animals vary from minor motor movements to generalized convulsions[11, 17]. Each animal has a stable seizure pattern so they can be grouped according to the type of seizure. The seizures are elicited by novelty, such as moving the animal out of the cage. The animals were studied over a year with a delay between the last seizure and sacrifice. Hippocampal neurons in H 3 were lost in the gerbils with generalized convulsion, although the maximum number of seizures was only 2 per/month over a year, which is a smaller number compared to the 150 (approximately) generalized convulsions elicited in the rats over a period of 2 months. The gerbils also had a reduction of Purkinje cells in relation to all types of seizures. These two experimental studies were planned meticulously so that ischemia should not be a problem as in status epilepticus. With regard to the natural seizures in the gerbils they are accompanied by interictal epileptic acitivity as a feature of the seizure prone condition. This is not the case in rats where seizures are only elicited by means of strong external stimulation.

The hypothesis which emerged from these studies is that the neuron loss could be related to the epileptic activity in itself. A study of epileptic activity in monkeys[34], with application of electronmicroscopy, revealed damage to selected neurons in hippocampus, that have been suggested to be inhibitory and GABA-ergic. This study supports the theory that seizure activity does damage neurons.

Discussion

Neuropathological research in epileptic patients has converged towards the area of most interest to the mechanism of epilepsy: whether hippocampal sclerosis is the cause or the effect of epilepsy. The field of neuropathology has developed in steps related to new investigative methods. The microscopic studies disclosed the rather restricted neuron loss in field H 1 in the hippocampus, which was named selective vulnerability. Many years passed before the phenomenon was reconfirmed in experimental research. Selective neuron loss was described in epilepsy and ischemia[35].

In the same research, different types of neuronal necrosis were previously described[21], but it is only recently that delayed neuron loss was observed. Most studies were acute studies with no delay before sacrifice. Delayed neuron loss occurs in particular in H 1 after a survival of the experimental animal of approximately 48 hours[15].

These newly recognized mechanisms of neuron loss may support the hypothesis that neuron loss in particular in H 1 might be due to hypoxia in early life, in accordance with theories based on human and experimental research. The neuron loss observed in epilepsy of man as well as from quantitative experimental research is most abundant in field H 3. The neuron loss here seems to be related to a sustained activity, probably of glutaminergic origin[35].

Neuron loss is a continuous danger in relation to epilepsy and not only with severe generalized convulsions. Whether kindling is involved in man is not known.

These results argue however for early epilepsy surgery, if possible and when necessary. However, the findings from neuropathology suggest that the surgical method should be reconsidered. The "en bloc" temporal lobe resection greatly facilitated research. However, it is now known that the best outcome is obtained when the neuron loss is restricted to the anterior part of the hippocampal structure. With proper selection the surgical approach might be restricted to the most anterior part of the hippocampus and amygdala.

References

1. Alzheimer A (1907) Die Gruppierung der Epilepsie. All Z Psychiat

2. Bouchet, Cazauvieilh (1825) Epilepsie et l'alienation mentale. Arch Gen Med 9: 510–542

3. Bratz F (1899) Ammonshorn Befunde der Epileptischen. Arch Psychiat Nervenkr 31: 820–936

4. Babb TL, Jann Brown W, Pretorius J, Davenport C, Lieb JP, Crandall PH (1984) Temporal lobe volumetric cell densities in temporal lobe epilepsy. Epilepsia 25: 729–740

5. Cocito L, Favale E, Reni L (1982) Epileptic seizures in cerebral arterial occlusive disease. Stroke 13: 189–196

6. Cooper IS (1973) Effect of chronic stimulation of anterior cerebellum on neurological disease. Lancet i: 206

7. Dam M (1972) The density and ultrastructure of the Purkinje cells following diphenylhydantoin treatment in animals and man. Acta Neurol Scand [Suppl] 49: 1–65

8. Falconer MA, Taylor DC (1968) Surgical treatment of drug-resistant epilepsy due to mesial temporal sclerosis. Arch Neurol 19: 353–361

9. Fritsch G, Hitzig E (1870) Über die electrische Erregbarkeit des Grosshirns. Arch Anat Physiol Med 37: 300

10. Gibbs EI, Gibbs EA, Fuster B (1948) Psychomotor epilepsy. Arch Neurol Psychiat 60: 331–339

11. Goddard GV, McIntyre DC, Leech CK (1969) A permanent change in brain function resulting from daily electrical stimulation. Exp Neurol 25: 295–330

12. Hughlings Jackson J, Stewart P (1899) Epileptic attacks with a warning of a crude sensation of smell and with the intellectual auro and symptoms of disorder in the right temporo-sphenoidal lobe. Brain 27: 550

13. Janz D (1964) Status epilepticus and frontal lobe lesions. J Neurol Sci 1: 446–457

14. Ketz E (1974) Brain tumours and epilepsy. In: Vinken P, Bruyn GW (eds) Handbook of clinical neurology, vol 16. North Holland Publ Comp, pp 254–269

15. Kirino T (1982) Delayed neuronal death in the gerbil hippocampus following ischemia. BrainRes 239: 57–69

16. Kopeloff LM (1960) Experimental epilepsy in the mouse. Proc Soc Exp Biol (N.Y.) 104: 500–504

17. Loskota WJ, Lomax P, Rich ST (1974) The gerbils as a model for the study of epilepsies. Epilepsia 15: 109–119

18. Lund M (1952) Epilepsy in association with intracranial tumour. Acta Neurol Psych Scand [Suppl] 81

19. Margerison JH, Corsellis JAN (1966) Epilepsy and the temporal lobes. Brain 89: 499–530

20. Meencke H-J, Janz D (1984) Neuropathological findings in primary generalized epilepsy: a study of eight cases. Epilepsia 25: 8–21

21. Meldrum BS (1983) Metabolic factors during prolonged seizures and their relation to nerve cell death. In: Delgado-Escueta AV et al (eds) Status epilepticus: Mechanisms of brain damage and treatment. Raven Press, New York, pp 261–276

22. Meynert T (1868) Studien über das pathologisch-anatomische Material der Wiener Irren-Anstalt. Vierteljahrzeitschr Psychiat 3: 381–402

23. Mouritzen Dam A (1982) Hippocampal neuron loss in epilepsy and after experimental seizures. Acta Neurol Scand 66: 601–642

24. Mouritzen Dam A, Fuglsang-Frederiksen A, Svarre-Olsen M, Dam M (1985) Late onset epilepsy: etiologies, types of seizure and value of clinical investigation, EEG and CT-scan. Epilepsia 26: 227–231

25. Mouritzen Dam A, Dam M (1986) Quantitative neuropathology in electrically induced generalized seizures. Conv Ther 2: 77–91

26. Penfield W, Erickson T (1941) Epilepsy and cerebral localization.Ch C Thomas, Springfield, Ill

27. Plum F, Posner JB, Troy B (1968) Cerebral metabolic and circulatory responses to induced convulsions in animals. Arch Neurol 18: 1–13

28. Rose M (1927) Der Allocortex bei Mensch und Tier. J Psychol Neurol 34: 1–111

29. Scholz W (1959) The contribution of pathoanatomical research to the problem of epilepsy. Epilepsia 1: 36–55

30. Serafetinides EA, Dominian J (1963) A follow-up study of late-onset epilepsy. I. Neurological findings. Br Med J 1: 428–431

31. Skyhøj Olsen T, Høgenhaven H, Thage O (1987) Epilepsy after stroke. Neurology 37: 1209–1211

32. Sommer W (1880) Erkrankung des Ammonshorns als ätiologisches Moment der Epilepsie. Arch Erkrank Nervenheilk 10: 631–675

33. Spielmeyer W (1927) Die Pathogenese des epileptischen Krampfes. Z Dtsch Ges Neurol Psychiatrie 109: 501–520

34. Söderfeldt B (1982) Epileptic brain damage. Histo-pathological studies on rats with experimental status epilepticus. Thesis, University of Lund, Sweden

35. Wieloch T, Lindvall O, Blomqvist P, Gage FH (1985) Evidence for amelioration of ischaemic neuronal damage in the hippocampal formation by lesions of the perforant path. Neurol Res 7: 24–26

Correspondence: Agnete Mouritzen Dam, MD, PhD, Neurological Research Laboratory, University Hospital Hvidovre Copenhagen, DK-2650 Hvidovre, Denmark.

Acta Neurochirurgica, Suppl. 50, 26–31 (1990)

Problems of Drug Surveillance – an International View

M. Massarotti[1] and **A. M. R. Matano**[2]

[1] Medical Department, Fidia S.p.A., Abano Terme (PD), Italy
[2] Clinical Drug Safety, Medical Department, Fidia S.p.A., Abano Terme (PD), Italy

Summary

The importance of Post Marketing Surveillance (PMS) is stressed, both to determine the therapeutic value of the drug and its adverse effects revealed in a large population of patients. A survey of such adverse effects of anti-epileptic drugs is provided: metabolic, dose – related and idiosyncratic. Problems of monotherapy and of drug interactions are discussed, as well as questions of central nervous system toxicity and of teratogenicity.

Keywords: Antiepileptic drugs; toxicity, surveillance.

Introduction

Every drug undergoes extensive testing in animals and humans before marketing, so approval for marketing is based on well-controlled clinical trials to demonstrate efficacy and safety, but the therapeutic value of the drug is determinated during Post-Marketing Surveillance (PMS). PMS can be viewed as a process that systematically monitors the patterns of use and the benefical or harmful effects of a drug as it is used in medical practice. PMS aims to resolve questions usually not answered during the more rigidly structured early studies.

Such questions include:

- definition of the optimum dose for the majority of patients;
- individualising treatment by defining sub-groups of patients that may respond in different ways or require different doses;
- definition of interactions with other drugs or diet;
- better surveillance of Adverse Drug Reactions (ADR): their incidence and their relationship to dosage and duration of therapy;
- comparison of ADR incidence and pattern with those of other treatments for the same condition;
- study of the response of patients to the drug under actual conditions of use and their compliance with treatment.

Hence Post-Marketing Surveillance protects the patients, allows the physician to use the drug better and enables pharmaceutical companies to develop and market more effective drugs.

Adverse Effects

Many adverse effects of antiepileptic drugs are known, but further unwanted effects of long-established drugs continue to emerge from time to time. Many interactions between antiepileptics, endogenous substances and other drugs have been recorded but much research has yet to be done to quantify their frequency and identify the mechanisms of the various interactions. The role of PMS is fundamental to this process.

A considerable amount of information is available concerning the clinical pharmacology of antiepileptic drugs[1–6]. The anticonvulsant drugs commonly used in clinical practice are: phenytoin (or its sodium salt), carbamazepine, phenobarbitone (and its congeners methylphenobarbitone and primidone), ethosuximide, valproic acid (or its sodium salt) and the benzodiazepines, clonazepam, nitrazepam and diazepam; other anticonvulsant drugs are used only occasionally, or have only recently been introduced. In order to select the most appropriate drug for the treatment of the epileptic patient, it is essential that the seizures are classified before the start of therapy; in fact, clinical experience has shown that certain anticonvulsant drugs tend to be most effective in certain types of epilepsy (Table 1).

Dosage

The appropriate dose of an antiepileptic drug is that which completely controls the epilepsy without causing

Table 1. *Correlation Between Types of Epilepsy and Effective Anticonvulsants*. (Modified from M. J. Eadie, 1984)

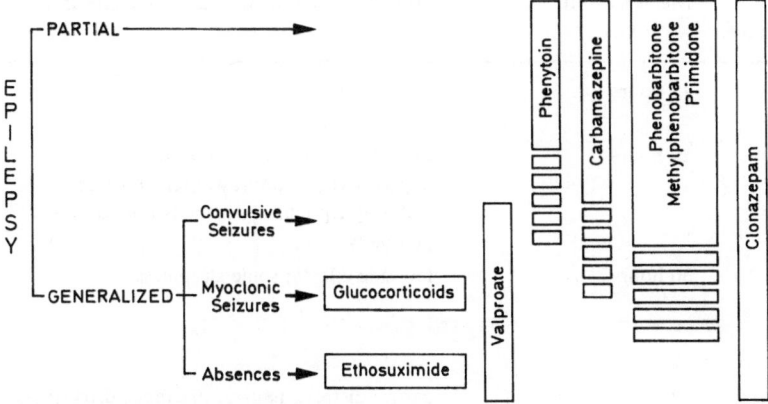

any unacceptable adverse effects. In order to determine appropriate drug treatment it is necessary to administer antiepileptic drugs in doses that produce steady-state plasma levels within the therapeutic range; in other words, to achieve an equilibrium between drug absorption and elimination[7] (Table 2). However, the therapeutic range relates to a given population but may need to be individualized for each patient. Some patients may experience toxic manifestations even when plasma levels are below the therapeutic range, particularly when polytherapy is used. The choice of an appropriate total daily dosage and the dosage schedule for each patient is not easy, because these may vary from patient to patient depending on age, presence of concomitant disease (hepatitis, renal failure), compliance, drug interactions, and differences in absorption, metabolism and excretion. Some of the adverse effects are shared by several of the drugs, others are peculiar to individual drugs. Antiepileptic drug toxicity is of three types: metabolic, dose-related and idiosyncratic.

Metabolic Toxicity

Metabolic toxicity may be manifested as folate deficiency, alterations in vitamin D metabolism (increased catabolism) or hypocalcaemia. In epileptic patients, low folic acid levels can be a result of malabsorption[8] or a consequence of antiepileptic therapy[9]. Many authors assert that phenytoin, phenobarbitone, primidone and carbamazepine may produce folate depletion if taken for a long period. However, folate depletion is unlikely to explain the antiepileptic effect that drugs have immediately after their initial administration. Prolonged drug treatment is possibly a cause of the slightly impaired ability of the liver to store vitamin B_{12}. Folate levels in CSF are two or three times those in serum,

even in folate deficiency. The blood-brain barrier may also play an effective role in protecting the central nervous system from the antifolate actions of a drug[10].

Folic acid is the preferred treatment for correcting megaloblastic anaemia in some patients on anticonvulsant therapy[11]. A significant reduction in serum calcium and phosphate and an increase in serum alkaline phosphatase may be found in up to 50% of adult patients treated with anticonvulsants, and alteration in vitamin D metabolism is generally thought to account for the hypocalcaemia and osteopenia. Hypocalcaemia and osteopenia occur in spite of normal levels of serum 25-hydroxycholecalciferol, suggesting that the hypocalcaemia is independent of the effect of the drugs on vitamin D metabolism[12].

Dose-related Toxicity

Dose-related side effects can be avoided by carefully monitoring plasma drug-levels and recognizing the potential for drug interactions which can occur if polytherapy is used. Dose-related side effects are the most common signs of toxicity of Carbamazepine. These include neurotoxic symptoms such as ataxia, diplopia, somnolence and coma. Carbamazepine can induce malaise if the full dose is initially administered. Patient's acceptance often is increased if these drugs are started slowly. However, maximum dosage should be determined by patent requirements and tolerability and not serum concentration. Benzodiazepines are promising newer agents for the chronic therapy of epilepsy but can cause excessive sedation when first used, but patients usually develop tolerance to them within a few weeks. The side effects of sodium valproate, such as gastrointestinal symptoms, hepatitis and pancreatitis,

Table 2. *Primary and Secondary Antiepileptic Drugs*

Drug	Usual adult dosage	Usual therapeutic range	Mean half-life	Some toxic and dose-related side effects
Carbamazepine (Tegretol®)	800–1,400 mg/day	6–12 g/ml	12 hours	nystagmus, ataxia, rash*
Clonazepam (Clonopin®)	highly individualized	5–50 g/ml	24 hours	drowsiness, exacerbation of childhood hyperactivity, withdrawal seizures and status epilepticus if dosage is not reduced gradually
Clorazepate (Tranxene®)	30–60 mg/day	metabolite desmethyl diazepam 1–2 g/ml	36 hours	few side effects; some sleepiness
Ethosuximide (Zarontin®)	1–2 g/day	40–100 g/ml	30 hours in children 60 hours in adults	gastric distress, nausea, dizziness, drowsiness
Mephenytoin (Mesantoin®)	400–700 mg/day	metabolites nirvanol plus mephenytoin 20–40 g/ml	4 days	drowsiness**
Methsuximide (Celontin®)	600–1,800 mg/day	metabolite desmethyl methsuximide 10–40 g/ml	40 hours	drowsiness, headache
Phenobarbital	60–300 mg/day	20–80 g/ml	4 days	mental changes; withdrawal seizures if stopped abruptly
Phensuximide (Milontin®)	1.5–3.0 g/day	unknown	4 hours	drowsiness, headache
Phenytoin (Dilantin®)	200–600 mg/day	10–30 g/ml	24 hours	mental dullness, ataxia, diplopia, gingival hypertrophy, acne, hirsutism
Primidone (Mysoline®)	500–1,500 mg/day	primidone 5–10 g/ml metabolites phenobarbital 10–30 g/ml	4 days	emotional and mental changes (including depression, irritability), impotence; withdrawal seizures if dosage is not reduced gradually
Trimethadione (Tridione®)	900–1,800 mg/day	unknown	18 hours	hemeralopia ("glare effect"); teratogenic
Valproic acid (Depakene®)	30–60 mg/kg/day	50–100 g/ml	8 hours	transient nausea, alopecia, fine tremor potential bleeding problems***

    * Blood dyscrasia; frequently check hematocrit and white blood cell count.
   ** High risk of blood dyscrasia; obtain hematocrit and white blood cell count monthly.
*** Liver damage; frequently check liver function.

appear more commonly at higher doses and at plasma concentrations over 100 mg/l[13].

Idiosyncratic Toxicity

Idiosyncratic effects are more difficult to avoid than dose-related effects, but physicians should be aware of the more common problems. Idiosyncratic toxicity can occur with any antiepileptic drug and can be fatal.

We would like to stress also that the toxic effects of these drugs may differ, depending on whether they are taken acutely or chronically. These factors are im-

portant since epilepsy is a common disorder affecting the young and the chronic and toxic effects have been exhaustively reviewed[9].

Monotherapy

Until a few years ago, the treatment of epilepsy often involved multiple drug therapy, but now a new concept has emerged, that of single-drug therapy. One of the factors that in the past has contributed to the use of polytherapy is the belief that the risk of toxicity was minimized by using small amounts of each antiepileptic drug; the concept of reduced toxicity applies only if the adverse effects are dose-related. Now, we know that many adverse reactions to anticonvulsants are not dose-related and the use of multiple antiepileptic drugs can increase the risk of toxicity. In fact a multiple drug regimen can have toxic effects even if the serum levels for each drug are within the therapeutic range.

For many patients single-drug therapy (monotherapy) results in better seizure control and fewer side effects than multiple-drug therapy (polytherapy)[14-21]. These workers assert that there is a correlation between a high frequency of seizure and the number of antiepileptic drugs administered. Polytherapy is used in cases where seizure disorders are more difficult to control[22].

Several reports have documented the benefits of monotherapy in adults, but few papers have discussed this matter in children[19,23]. For example, in children, the risk of severe (fatal) hepatotoxicity caused by valproate is considerably lower with valproate monotherapy (1 : 10,000 patients) than with polytherapy[24].

In addition, antiepileptic drugs interact[25,26] frequently making it almost impossible to obtain therapeutic concentrations of one drug in plasma when it is being given in combination with another[23]. In any case, much higher doses of each antiepileptic drug must be administered to maintain adequate plasma concentrations during polytherapy than monotherapy[27]. The concept of a therapeutic range is confounded when the patient is taking more than one anticonvulsant.

Drug Interactions

A large number of interactions between antiepileptic drugs, or between these and other drugs, are known[28,30]. The reported interactions are too numerous to be considered in detail here. However, the principal concept to be stressed is that the enzyme-inducing anticonvulsants have been documented to increase the metabolism of other concomitantly administered lipid-soluble drugs. Further research is needed to determine the risk-benefit ratio of switching from poly-to monotherapy. Nevertheless, efforts must be made to learn whether each child with epilepsy can benefit from a program of monotherapy.

Central Nervous System Toxicity

Another big problem relating to these drugs is their toxic effect on the central nervous system; in fact, epilepsy is frequently associated with cognitive dysfunction[31]. Authors have suggested that intellectual function in patients with epilepsy is influenced by a combination of factors such as: age, duration of epilepsy, frequency and type of seizures, aetiology of seizures, hereditary factors, psychosocial issues and antiepileptic drug therapy[32].

Anticonvulsant drugs appear to differ in their cognitive effects. In 1985 the Committee on Drugs of the American Academy of Pediatrics evaluated the evidence concerning the cognitive effects of the major antiepileptic drugs[33]. The findings are summarized below.

(a) Phenobarbital: selectively impairs short-term memory; memory concentration and symbol manipulation also are affected.

(b) Phenytoin: affects attention, performance on neuropsychological tests of problem solving and visuomotor tasks.

(c) Carbamazepine: may have some adverse effects on task performance (effect concentration).

(d) Valproate: has shown no effect on learning tasks (minimal adverse effects on cognition).

Recent studies indicate that a reduction in polytherapy is often accompanied by neuropsychological improvement[34]. Clinical evidence suggests that adverse cognitive drug effects may be reversible in many cases when therapy is simplified or discontinued[35,36]. This evidence is very important particularly for children. Reversibility of adverse cognitive drug effects could have implications for long-term intellectual development.

Teratogenicity

It is known that the incidence of foetal malformations is increased (approximately doubled or trebled) in infants of mothers taking antiepileptics[37,38]. In recent prospective studies teratogenicity appears to be attributable to antiepileptic drugs (mainly phenytoin, phenobarbitone or primidone) rather than to the epilepsy; in fact, more congential anomalies have been found among the newborn of treated epileptic mothers than among untreated epileptic mothers[39-43]. But it has been observed that when phenobarbitone is taken in pregnancy, though not for epilepsy, there is no increase in foetal abnormalities. However, there is an increase if the drug is taken for maternal epilepsy[37]. These data emphasize that there are many factors that may contribute to the production of malformations in the newborn, such as: genetic history[44], harmful effects of drug metabolites[45,46], drug interference with folate metabolism, depression of cardiorespiratory functions and foetal hypoxia due to maternal seizures[47].

Recent studies have demonstrated that drug absorption, protein-binding and clearance are altered during pregnancy. In fact, serum levels of antiepileptic drugs tend to decrease and therefore a change in dosage may be necessary to maintain plasma anticonvulsant concentrations, particularly in the last trimester[39, 48, 49]. The precise mechanisms involved in the increased clearance with pregnancy are not yet clear. There was no definite dose-dependent increase in the incidence of dysmorphogenesis associated with any individual antiepileptic drug. Also, there was no relationship between the type of defect and the individual antiepileptic drug[38, 50]. Moreover, there was no significant relationship between type of maternal epilepsy (seizure type) and the occurence of malformation[50]. The rate of malformation is related to the number of drugs to which the infant was exposed; the role of polypharmacy in teratogenesis has been discussed by many physicians[42, 48, 50, 51]. Analysis of the literature reveals that many studies on the teratogenic effects of antiepileptic drugs have been conducted but no definite conclusions have yet been reached.

Finally, it is necessary to keep under continuous review which side effects are "acceptable" in order to provide the best possible quality of life for the patient.

References

1. Eadie MJ, Tyrer JH (1980) Anticonvulsant therapy: pharmacological basis and practice, 2nd ed. Churchill Livingstone, London Edinburgh New York
2. Hvidberg EF, Dam M (1976) Clinical pharmacokinetics of anticonvulsant. Clin Pharmacokinet 1: 161–188
3. Johannessen SI (1981) Antiepileptic drugs: pharmacokinetic and clinical aspects. Ther Drug Monit 3: 17–37
4. Richens A, Scoular IT, Ahmad S, Jordan BJ (1976) Pharmacokinetics and efficacy of Epilim in patients receiving longterm therapy with other anticonvulsants. In: Legg (ed) Clinical and pharmacological aspects of sodium valproate (Epilim) in the treatment of epilepsy. MCS Consultants, Tunbridge Wells, pp 78–88
5. Vida JA (1977) Anticonvulsants. Academic Press, New York
6. Woodbury DM, Penry JK, Pippenger CE (1982) Antiepileptic drugs, 2nd ed. Raven Press, New York
7. Moretti Ojemann L, Ojemann GA (1984) Treatment of epilepsy. AFP 30: (2), 113–128
8. Ibbotson RN, Dilena BA, Hordwood JM (1967) Studies on deficiency and absorption of folates in patients on anticonvulsant drugs. Aust Ann Med 16: 144–150
9. Reynolds EH (1975) Chronic antiepileptic toxicity: a review. Epilepsia 16: 319–352
10. Dastur DK, Dave UP (1987) Effect of prolonged anticonvulsant medication in epileptic patients: serum lipids, vitamins B_6, B_{12} and folic acid, proteins and fine structure of liver. Epilepsia 28 (2): 147–159
11. Martindale J (1982) Phenytoin and some other convulsants. In: Reynolds ER (ed) The extrapharmacopoeia. Pharmaceutical Press, London, 1235 p
12. Davis-Jones GAB (1986) Anticonvulsant drugs. In: Dukes MNG (ed) Side effects of drugs annual 10. Elsevier Science Publishers B.V. Amsterdam, pp 53–57
13. Brodie MJ (1985) The optimum use of anticonvulsants. The Practitioner 229: (1408), 921–927
14. Shakir RA, Johnson RH, Lambie DG et al. (1981) Comparison of sodium valproate and phenytoin as single drug treatment in epilepsy. Epilepsia 22: 227–233
15. Strandjord RE, Johannessen SJ (1980) Single drug therapy with carbamazepine in patients with epilepsy: serum levels and clinical effect. Epilepsia 2: 662–665
16. Feely MP, Callaghan N (1981) Changing to single drug therapy in epilepsy. Lancet i: 847
17. Fischbacher E (1982) Effect of reduction of anticonvulsants on wellbeing. Br Med J 285: 423–424
18. Maheshwari MC, Padmini R (1981) Role of carbamazepine in reducing polypharmacy in epilepsy. Acta Neurol Scand 64: 22–28
19. Cloyd JC, Kriel RL, Fischer JH (1985) Valproic acid pharmacokinetics in children. II. Discontinuation of concomitant drug therapy. Neurology 35: 1623–1627
20. Callaghan N, O'Dwyer R, Keating J (1984) Unnecessary polypharmacy in patients with frequent seizures. Acta Neurol Scand 69: 15–19
21. Bennet HS, Dunlop T, Ziring P (1983) Reduction of polypharmacy for epilepsy in an institution for the retarded. Dev Med Child Neurol 25: 735–737
22. Beran G, Sutton C (1982) Treatment of epilepsy: monotherapy versus polytherapy. Med J Aust (2): 135–138
23. Reynolds EH, Shorvon SD (1981) Monotherapy or polytherapy for epilepsy. Epilepsia 22: 1–10
24. Dreifuss FE, Santilli N, Langer DH, Sweeny KP, Moline KA, Menander KB (1987) Valproic acid hepatic fatalities: a retrospective review. Neurology 37: 379–385
25. Bowdle TA, Levy RH, Cutler RE (1979) Effects of carbamazepine on valproic acid kinetics in normal subjects. Clin Pharm Ther 26: 629–634
26. Bruni J, Gallo JM, Lee CS et al (1980) Interactions of valproic acid with phenytoin. Neurology 30: 1233–1236
27. Henriksen O, Johannessen SI (1982) Clinical and pharmacokinetic abservations on sodium valproate – a 5-year follow-up study in 100 children with epilepsy. Acta Neurol Scand 65: 504–523
28. Richens A (1977) Interaction with antiepileptic drugs. Drugs 13: 266–275
29. Perucca E, Richens A (1980) Anticonvulsant drug interactions. In: Tyrer JH (ed) The treatment of epilepsy. Lancaster MTP Press, pp 95–128
30. Perucca E (1982) Pharmacokinetic interactions with antiepileptic drugs. Clin Pharmacokinet 7: 57–84
31. Hirtz DG, Nelson KB (1985) Cognitive effects of antiepileptic drugs. In: Pedley TA, Meldrum BS (eds) Recent advances in epilepsy. Churchill Livingstone, New York, pp 161–181
32. Lesser RP, Luders H, Wyllie E, Dinner DS, Morris HH III (1986) Mental deterioration in epilepsy. Epilepsia 27: [Suppl 2] S 105–23
33. Committee on Drugs, American Academy of Pediatrics (1985) Behavioral and cognitive effects of anticonvulsant therapy. Pediatrics 76: 644–647

34. Reynolds EH, Shorvon SD (1981) Monotherapy or polytherapy for epilepsy? Epilepsia 22: 1–10

35. Trimble MR, Thompson PJ (1983) Anticonvulsant drugs, cognitive function, and behavior. Epilepsia 24: (S 1) S 555–63

36. Vining EPG, Shinnar SS, Mellits D, Silverton S, Brandt J (1986) Discontinuing antiepileptic drugs improves intellectual function (abstract). Epilepsia 27: 639

37. Shapiro S, Hartz SC, Siskind V, Mitchell AC, Slone D, Rosenberg L, Monson RR, Heinonen OP, Idanpaan-Heikkila J, Haro S, Saxén L (1976) Anticonvulsants and parental epilepsy in the development of birth defects. Lancet 1: 272–275

38. Janz D, Bossi L, Dam M, Helge H, Richens A, Schmidt D (1982) Epilepsy, pregnancy and the child. Raven Press, New York

39. Bossi L, Battino D, Boldi B, Caccamo HL, Ferraris G, Latis GO, Simionato L (1982) Anthropometric data and minor malformations in newborns of epileptic mothers. In: Janz D, Bossi L, Dam M, Helge H, Richens A, Schmidt S (eds) Epilepsy, pregnancy and the child. Raven Press, New York, pp 299–301

40. Dansky L, Andermann F (1982) Major congential malformations in the offspring of epileptic parents: genetic and environmental risk factors. In: Janz D, Bossi L, Dam M, Helge H, Richens A, Schmidt S (eds) Epilepsy, pregnancy and the child. Raven Press, New Nork, pp 223–34

41. Granstrom ML, Hiilesmaa VK (1982) Malformations and minor anomalies in the children of epileptic mothers: preliminary results of the prospective Helsinki study. In: Janz D, Bossi L, Dam M, Helge H, Richens A, Schmidt S (eds) Epilepsy, pregnancy and the child. Raven Press, New York, pp 303–7

42. Nakane Y (1982) Factors influencing the risk of malformation among infants of epileptic mothers. In: Janz D, Bossi L, Dam M, Helge H, Richens A, Schmidt S (eds) Epilepsy, pregnancy and the child. Raven Press, New York, pp 259–65

43. Kaneko S, Fukushima Y, Sato T, Nomura Y, Ogawa Y, Saito M, Shinagawa S, Yamazaki S (1984) Hazards of fetal exposure to antiepileptic drugs: a preliminary report. In: Sato T, Shinagawa S (eds) Antiepileptic drugs and pregnancy. Excerpta Medica, Amsterdam, pp 132–138

44. Koch S, Hartmann AM, Jager-Roman E, Rating D, Helge H (1982) Major malformations in children of epileptic parents – due to epilepsy or its therapy? In: Janz D, Bossi L, Dam M, Helge H, Richens A, Schmidt S (eds) Epilepsy, pregnancy and the child. Raven Press, New York, pp 313–315

45. Lindhout D, Hoppener R, Meinardi H (1984) Teratogenicity of antiepileptic drug combinations with special emphasis on epoxidation (of carbamazepine). Epilepsia 25: 77–83

46. Nau H, Loscher W (1984) Valproic acid and metabolites: pharmacological and toxicological studies. Epilepsia 25: S 14–22

47. Millicovsky G (1981) Maternal hyperoxia greatly reduces the incidence of phenytoin-induced cleft lip and palate in A/J mice. Science 212: 671–72

48. Nakane Y, Okuma T, Takahashi R, Sato Y, Wada T, Sato T, Fukushima Y, Kumashiro H, Inami M, Komai S, Seino M, Otsuki S, Hosokawa K, Inanaga K, Nakazawa Y, Yamamoto K (1980) Multi-institutional study on the teratogenicity and fetal toxicity of antiepileptic drugs: a report of a collaborative study group in Japan. Epilepsia 21: 663–80

49. Nau H, Kuhnz W, Egger H-J, Rating D, Helge H (1982) Anticonvulsant during pregnancy and lactation: transplacental, maternal and neonatal pharmacokinetics. Clin Pharmacokinet 7: 508–543

50. Kaneko S, Otani K, Fukushima Y, Ogawa Y, Nomura Y, Ono T, Nakane Y, Teranishi T, Goto M (1988) Teratogenicity of antiepileptic drugs: analysis of possible risk factors. Epilepsia 29: (4). 459–67

51. Rating D, Jager-Roman E, Koch S, Gopfert-Geyer I, Helge H (1982) Minor anomalies in the offspring of epileptic parents. In: Janz D, Bossi L, Dam M, Helge H, Richens A, Schmidt S (eds) Epilepsy, pregnancy and the child. Raven Press, New York, pp 283–8

52. Eadie MJ (1984) Anticonvulsant drugs an update. Drugs 27: 328–363

Correspondence: M. Massarotti, M.D., Medical Department, Fidia S.p.A., Via Ponte della Fabbrica, 3/A, I-35031 Abano Terme, Italy.

Acta Neurochirurgica, Suppl. 50, 32–37 (1990)

The Epidemiology of Drug Resistant Epilepsy and Adverse Effects of Antiepileptic Drugs

D. Chadwick

Regional Neurological Center, Walton Hospital, Liverpool, U.K.

Summary

Patients with partial, particularly complex partial, epilepsy, especially where this is related to underlying cerebral disease or damage, tend to respond poorly to existing antiepileptic drug therapy. The epidemiology of such patients is reviewed, together with the adverse effects of antiepileptic drug therapy which may be acute (dose-related and idiosyncratic) or chronic. Such chronic toxicity may cause nervous system, skin, hepatic, haematological, endocrine, and connective tissue problems, and also disorders of pregnancy. As complex partial epilepsy often responds poorly to drug therapy, the possible benefit of surgical treatment should be considered at a relatively early stage.

Keywords: Epilepsy; drug resistance; epidemiology; antiepileptic drugs; adverse effects.

The Epidemiology and Prognosis of Epilepsy

Epilepsy is a disorder resulting from hyper-excitability leading to abnormal discharge of cortical neurons. As such, it is a heterogenous disorder with a multiplicity of causes and very varied prognosis. Epileptic seizures and their phenomenology are similarly complex. The purpose of this brief review is to place into context those epilepsies that may be treatable by surgical means and to consider the effects of antiepileptic drug therapy in such patients. It is clear that they constitute a minority of patients who develop epilepsy at some time in their lives.

Approximately 5/1000 of the population have epilepsy at any single point in time[18]. However, age specific prevalence and incidence rates vary greatly (Fig. 1), with an overall incidence of approximately 50/100.000. For new cases of epilepsy it appears that between a half and two thirds of these patients will have seizures which arise focally[8]. Of those with focal or partial epilepsies approximately 60% will have complex partial seizures.

Historically, the medical profession's view of the prognosis of epilepsy has been overly pessimistic. Community based studies of the prognosis of epilepsy would indicate that up to 70 or 80% of patients attain long term remissions of their epilepsy and perhaps 50% of patients developing epilepsy achieve long-term remissions which are independent of antiepileptic drug therapy[3]. This finding is supported by more recent studies of populations of patients presenting with epilepsy for the first time which show similarly high remission rates very shortly after the commencement of therapy[16, 21]. A number of factors clearly do predict the likelihood of remission. Adverse factors are the presence of partial rather than primary generalised seizures, the presence of underlying cerebral disease resulting in a symptomatic epilepsy, and the long duration and high frequency of seizures and possibly the number of seizures before treatment is commenced[15]. Thus, patients with chronic intractable epilepsy, poorly responsive to existing antiepileptic drug therapy, tend to be those patients with partial, particularly complex partial epilepsy, especially where this is related to some underlying cerebral disease or damage. Even a brief attendance at any neurological follow up clinic or epilepsy clinic will be sufficient to convince the observer that it is this group of patients that represent the greatest therapeutic challenge.

Complex Partial Epilepsy

In complex partial epilepsy, complex partial seizures occur with or without secondary generalised tonic-clonic seizures. The response to antiepileptic drug treatment is poor with probably less than 50% of patients achieving long term remissions[21]. Very often treatment with antiepileptic drugs will prevent secondary generalised seizures but the complex partial seizures remain.

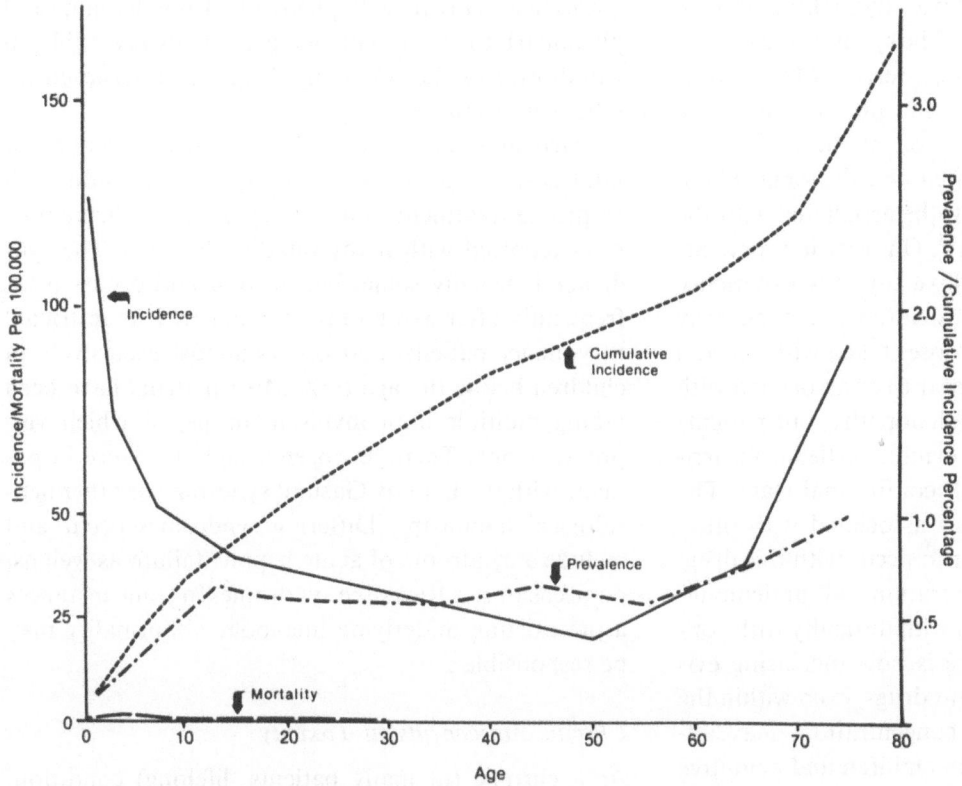

Fig. 1. Age specific incidence and prevalence rates for epilepsy in Rochester, Minnesota 1935–1974. (From Anderson *et al* 1986, with permission)

Because of the association of altered state of awareness, automatism and often prolonged post-ictal confusion, such seizures represent a very significant remaining disability for many patients with this epilepsy. The epilepsy also presents difficulties because of its strong association with a variety of psychiatric disorders. Considerable debate exists as to whether particular personality disorders can be associated with temporal lobe epilepsy[1], but there can be no doubt that affective disorders are common in temporal lobe epilepsy and that occasionally schizophreniform psychoses will occur.

The causes of complex partial epilepsy often remain obscure. However, a recent community-based case-control study of 82 patients with complex partial epilepsy has shown that as many as 20% of patients may have such an epilepsy related to prolonged complicated febrile seizures in childhood[17]. Other important causes in this study were head injury and perinatal insult.

Can we then define the outlook for patients with complex partial epilepsy who do not enter early remission on treatment with a single drug? There is no very convincing evidence that any one antiepileptic drug is clearly superior in its efficacy against partial seizures than any other[4]. Substitution of one drug for another is therefore rarely successful. The addition of a second drug to a first may result in long-term re-

mission or improvement for a further 10–15% of patients[12, 20]. In spite of the poor chance of significant improvement in seizure control with additional drugs, it is very commonly seen that patients with intractable complex partial epilepsy pass with time onto a regime that may include 2, 3 or more antiepileptic drugs. Thus, whilst there is little evidence that such increases in treatment have a great effect on control of epilepsy there is nevertheless abundant evidence that such practices increase the risks of antiepileptic drug toxicity[13].

Adverse Effects of Antiepileptic Drugs

Complications of anticonvulsant therapy may be classified into three differing types:
1. Acute dose-related side effects
2. Acute idiosyncratic side effects
3. Chronic toxicity.

Acute Dose-related Side Effects

Anticonvulsants, including phenytoin, carbamazepine, phenobarbitone, primidone and benzodiazepines, give rise to a non-specific encephalopathy associated with high blood concentrations of the drug concerned[19]. Patients initially present with increased tiredness and show a gaze-paretic nystagmus. With increasing serum levels the patients become ataxic, dysarthric, and ultimately confused and drowsy. The syndrome is usually associated with the presence of asterixis. More rarely, patients (particularly those with pre-existing brain damage) may exhibit a variety of involuntary movement disorders including chorea and dystonia. Phenytoin is the drug most likely to be associated with the development of such reactions[5]. On occasion, patients with high serum levels parodoxically may experience an exacerbation of their seizures. Status epilepticus may occur as a result of large overdoses taken with suicidal intent. Valproate does not appear to be associated with a typical anticonvulsant encephalopathy, but patients with high serum levels may exhibit restlessness, irritability and, at times, a frank confusional state. The cerebellar signs and symptoms associated with other anticonvulsants do not appear to occur with this drug.

While intoxicating concentrations of anticonvulsants are inevitably associated with difficulty with concentration and memory, there is now increasing evidence that many anticonvulsant drugs, even within the therapeutic range of serum concentration, may adversely affect both behavior in children and cognitive function. These effects seem to be greatest with phenytoin and barbiturate drugs, and are less marked with newer drugs such as carbamazepine and valproate[14].

Acute Idiodyncratic Side-effects

The commonest acute idiosyncratic reaction to anticonvulsant drugs is a maculopapular exanthematous eruption, which usually occurs within a month of initiating treatment[19]. Such reactions are commonly seen with phenytoin (incidence up to 10%) and carbamazepine (incidence up to 15%). High initial serum concentrations of these anticonvulsants may increase the risk of such skin eruptions[6]. More rarely, more severe skin eruptions may occur which include exfoliative dermatitis and the Stevens-Johnson syndrome.

Aplastic anaemia occurs very rarely with most anticonvulsant drugs, but has attracted particular attention in the case of carbamazepine. In fact, the association between carbamazepine and aplastic anaemia in patients with epilepsy remains extremely rare and most of the initial reports were related to use of the drug in elderly patients who were suffering from trigeminal neuralgia. Aplastic anaemia has also been reported with phenytoin, ethosuximide and primidone.

An acute hepatitis may occur with a number of anticonvulsant drugs. It is usually associated with fever, lymphadenopathy and skin eruption and has been recorded as occurring with phenytoin, trimethadione and phenobarbitone. The changes are usually reversible on withdrawal of the offending drug, but occasional fatalities do occur.

Particular concern has arisen because of reports of fatal cases of acute liver failure in association with valproate treatment. Some 40 or more case have now been reported with many puzzling features. The syndrome is usually somewhat delayed and occurs most frequently after 3–6 months treatment. It is restricted to younger patients and occurs almost exclusively in children below the age of 2. Most patients have been taking multiple anticonvulsant drugs, of which valproate is one. There has been a high incidence in patients with the Lennox Gastaut syndrome or other neurological handicaps. Differing syndromes occur and include a syndrome of acute hepatic failure as well as, on occasions, a Reye-like syndrome. In some instances a pre-existing underlying metabolic abnormality may be responsible[7].

Chronic Anticonvulsant Toxicity

As a chronic (in many patients, lifelong) condition, epilepsy is unusual in that drug treatment may be necessary for several decades of a patient's life. Over the years, a wide range of chronic toxic effects have been recognised. Many are subtle, but on occasions they may become clinically important. It seems that the incidence of chronic toxicity is likely to increase with increasing doses of anticonvulsants and with multiple drug therapy. Indeed, because of the common practice of treating epileptic patients with multiple drugs, it is at times difficult to attribute a given symptom of toxicity to any one specific drug. The chronic toxic side effects of anticonvulsants are summarized in Table 1.

Nervous System Toxicity

An unusual syndrome of pseudodementia, declining cognitive function and increasingly frequent seizures may be seen in younger mentally retarded patients receiving phenytoin. These patients usually have intoxicating serum concentrations of the drug.

It has long been recognised that there is a high incidence of cerebellar atrophy at postmortem in epileptic patients. Histologically this is associated with loss of Purkinje cells.

A mild, mainly sensory neuropathy, usually with loss of lower limb reflexes, may occur in up to 10% of patients receiving phenytoin over long periods. Usually such patients are receiving other anticonvulsant drugs.

Table 1. *Chronic Toxicity of Anticonvulsants*

Nervous system
  Memory and cognitive impairment
  Hyperactivity and behavioural disturbance
  Pseudodementia
  Cerebellar atrophy
  Peripheral neuropathy

Skin
  Acne
  Hirsutism
  Alopecia
  Chloasma

Liver
  Enzyme induction

Blood
  Megaloblastic anaemia
  Thrombocytopenia
  Lymphoma

Immune system
  IgA deficiency
  Drug-induced SLE

Endocrine system
  Decreased thyroxine levels
  Increased cortisol and sex hormone metabolism

Bone
  Osteomalacia

Connective tissue
  Gum hypertrophy
  Coarsened facial features
  Dupuytren's contracture

Pregnancy
  Obstetric complications
  Teratogenicity
  Fetal hydantoin syndrome

A variety of behavioual changes may be seen with anticonvulsant drugs: in particular, phenobarbitone is well recognised to cause a syndrome of hyperacitivity and behavioural disturbance in children.

Skin Disorders

Acne and hirsutism are commonly seen following treatment with phenytoin and possibly also with barbiturates. These effects are so troublesome that it would seem unreasonable to treat young girls with phenytoin if this can be avoided in any way. Valproate may on occasion be associated with alopecia, which is usually temporary even when the drug is continued.

Hepatic Metabolism

Some clinical changes in enzymes occur very commonly in anticonvulsant-treated patients. Gamma-glutamyl-transpeptidase may be increased in a high percentage of patients, but changes in plasma alkaline phosphatase and aspartate transaminase levels are less frequently seen.

Many anticonvulsant drugs (most notably phenytoin, carbamazepine, and barbiturates) are potent liver-enzyme inducers. This increased microsomal liver enzyme acitivity may contribute to many other aspects of chronic anticonvulsant toxicity.

Haematological Disorders

There is a well-documented trend for mean cell volume to rise following the initiation of anticonvulsant treatment, usually with phenytoin. Frank folate deficiency may develop and lead to a megaloblastic anaemia. The manner in which phenytoin leads to folate deficiency remains uncertain, but enzyme induction may be important. Sodium valproate may occasionally be associated with a thrombocytopenia, which (very rarely) may result in bleeding disorders in children.

There is an increased incidence of lymphoma in patients treated for epilepsy, which may be two or three times the expected incidence.

Endocrine System Disorders

It is well recognised that anticonvulsant-treated patients have lowered plasma thyroxine levels. This fact is not clinically significant and, like the increased excretion of steroid metabolites in epileptic patients, is probably due to hepatic enzyme induction. The latter has clinical significance in causing an increased incidence of breakthrough bleeding and contraceptive failure in anticonvulsant-treated patients taking oral contraceptive preparations. Increased metabolism of vitamin D is probably responsible for anticonvulsant-induced osteomalacia which is sometimes seen in patients with severe epilepsy.

Connective Tissue Changes

Gum hypertrophy may occur in up to 50% of patients treated with phenytoin and is exacerbated by poor dental hygiene. Many epileptic patients, usually those with severe epilepsy and receiving multiple drugs, develop coarsened facial features with thickening of the lips, widened nose and thickening of subcutaneous facial tissue. Many of these changes may be drug related, but a similar epileptic facies was also documented before the advent of modern anticonvulsant therapy.

Disorders of Pregnancy

Studies show that epileptic women suffer a higher rate of obstetric complications including antipartum haemorrhage, premature labour and a higher rate of caesarian sections. The children of epileptic mothers tend to have lower birth weights and smaller head circumference than the expected.

Children of epileptic mothers show an increased rate of major foetal abnormality (between two and three times the expected)[9]. These most commonly take the form of hare lip, cleft palate abnormalities and cardiovascular anomalies, usually associated with phenytoin and older anticonvulsant drugs. A milder pattern of foetal abnormality has been associated particularly with phenytoin treatment during pregnancy: this includes facial abnormalities with hypertelorism, flattened bridge to the nose and epicanthic folds with low-set ears and wide mouth. These changes may be associated with a retarded growth and a degree of mental retardation. Recently, it has been suggested that valproate is associated with an increased incidence of spina bifida abnormalities (approximately 1–2% of pregnancies: two to three times the expected rate)[11].

Although the increased incidence of these complications in pregnancy is partly related to drug treatment, other factors are almost certainly involved: these include genetic associations between epilepsy and foetal abnormalities, as well as complications to the pregnancy and the child resulting from seizures during pregnancy. For this reason, while it is sensible to withdraw anticonvulsant therapy before a pregnancy in women with prolonged remissions in their epilepsy, most patients who are clearly dependent on anticonvulsant drugs to control their epilepsy should be maintained on anticonvulsant therapy during their pregnancy. In the case of valproate, screening for spina bifida with ultrasound, amniocentesis and α-fetoprotein should be undertaken at an early stage of pregnancy.

Conclusion

Making estimates from what is known of the incidence and prevalence of epilepsy in general and complex partial epilepsy in particular, it has been calculated that as many as 1.7 per hundred thousand of the population may benefit from surgical treatment of complex partial epilepsy each year[10]. It is clear that for a group of patients with complex partial epilepsy resistant to an-

tiepileptic drugs, continued life-long therapy holds relatively little hope of attaining remission but does carry with it significant risks of adverse reactions. As it seems that the long-term prognosis of epilepsy can be predicted very quickly after the onset of treatment[15] there seems no doubt that many more patients need to be considered for surgical treatment than is currently the case. This conclusion has major implications for the provision of resources to allow adequate investigation and selection of patients for surgical treatment of epilepsy.

References

1. Aird RB, Masland RL, Woodbury DM (1985) The epilepsies: a critical review. Raven Press, New York, pp 158–180
2. Anderson VE, Hauser WA, Rich SS (1986) Genetic heterogeneity in the epilepsies. Advances in Neurology, Vol 44. Raven Press, New York, pp 59–75
3. Annegers JF, Hauser WA, Elverback LR (1979) Remission of seizures and relapse in patients with epilepsy. Epilepsia 20: 729–737
4. Chadwick D, Turnbull MD (1985) The comparative efficacy of antiepileptic drugs for partial and tonic-clonic seizures. Neurol Neurosurg Psychiatry 48: 1073–1077
5. Chadwick D, Reynolds EH, Marsden CD (1976) Anticonvulsant-induced dyskinesias: A comparison with dyskinesias induced by neuroleptics. J Neurol Neurosurg Psychiatry 39: 1210–1218
6. Chadwick D, Shaw MDM, Foy P, Rawlins MD, Turnbull DM (1984) Serum anticonvulsant concentrations and the risk of drug induced skin eruptions. J Neurol Neurosurg Psychiatry 47: 642–644
7. Dreifuss FE et al (1987) Valproic acid fatalities: a retrospective review. Neurology 37: 379–385
8. Hauser WA, Kurland LT (1975) Epidemiology of epilepsy in Rochester, Minnesota, 1935–1967. Epilepsia 16: 1–66
9. Janz D (1982) On major malformations and minor anomalies in the offspring of patients with epilepsy: review of the literature. In: Janz D et al (ed) Epilepsy, pregnancy and the child. Raven Press, New York, pp 211–222
10. Janz D (1987) Consequence for the practice of epilepsy therapy in Europe. In: Wieser HG, Elger CE (eds) Presurgical evaluation of epileptics. Springer, Berlin Heidelberg New York, pp 373–376
11. Lindhout D, Schmidt D (1986) In-utero exposure to valproate and neural tube defects. Lancet 1392–1393
12. Mattson RH et al (1985) Comparison of carbamazepine, phenobarbital, phenytoin and primidone in partial and secondary generalised tonic-clonic seizures. New England J Med 313: 145–151
13. Reynolds EH (1975) Chronic antiepileptic toxicity: a review. Epilepsia 16: 319–352
14. Reynolds EH (1983) Mental effects of antiepileptic medication: a review. Epilepsia 24: 585–595
15. Reynolds EH, Elwes RDC, Shorvon SD (1983) Why does epilepsy become intractable? Prevention of chronic epilepsy. Lancet ii: 952–954

16. Reynolds EH, Shorvon SD (1981) Monotherapy or polytherapy for epilepsy? Epilepsia 22: 1–10

17. Rocca WA, Sharbrough FW, Hauser WA, Anneger JF, Schoenberg BS (1987) Risk factors for complex-partial seizures: a population-based case-control study. Ann Neurol 21: 22–31

18. Sander JWAS, Shorvon SD (1987) Incidence and prevalence studies in epilepsy and their methodological problems: a review. J Neurol Neurosurg Psychiatry 50: 829–839

19. Schmidt D (1982) Adverse affects of antiepileptic drugs. Raven Press, New York

20. Schmidt D (1982) Two antiepileptic drugs for intractable epilepsy with complex-partial seizures. J Neurol Neurosurg Psychiatry 45: 1119–1124

21. Turnbull DM, Howell D, Rawlins MD, Chadwick D (1985) Which drug for the adult epileptic patient: phenytoin or valproate? Br Med J 290: 815–819

Correspondence: D. Chadwick, M.D., Regional Neurological Center, Walton Hospital, Rice Lane, Liverpool L9 1AE, U.K.

Acta Neurochirurgica, Suppl. 50, 38–47 (1990)

Posttraumatic Epilepsy. Incidence and Prophylaxis

C. A. Pagni

2nd Chair of Neurosurgery, University of Torino, Italy

Summary

A detailed review, based on the literature and the author's own series, is given of the incidence of both early and late epilepsy following head injury related to age, severity and other specific features of the injury and clinical sequelae. Use of prophylactic anti-convulsant therapy, following head injury, remains controversial despite positive results of animal experiments. Hence, the author recommends that antiepileptic medication should be restricted to patients who have had at least two epileptic fits during the first two years after injury.

Keywords: Epilepsy; posttraumatic; incidence; prophylaxis.

1. Introduction

The overall incidence of posttraumatic epilepsy in different series of consecutive, unselected trivial and severe injuries, including mainly closed head injuries, is about 3–5%. If a large number of open head injuries is included the incidence rises to 8–9% (Table 1).

In the series of non-missile combat injuries the incidence reaches values of 12–24%; in penetrating missile injuries the incidence is 34–53% (Table 2).

In non-missile civilian brain wounds the occurrence of epilepsy is superimposable on that of penetrating missile injuries (Table 3).

Epilepsy may occur either soon after injury or later. Fits occurring immediately or within a few hours or days after injury have been called "early fits"; those occurring weeks or months after injury "late fits". There was no general consensus about the definition of "early epilepsy" and authors have variously referred

Table 1. *Overall Incidence of Posttraumatic Epilepsy*

	Year	Total	With epilepsy
Jennett	1975	1,000	5%
Paillas *et al.*	1970	2,145	188 = 9%
De Santis and Pagni	1976 child	2,546	139 = 5.45%
De Santis *et al.*	1979 adult	2,980	84 = 2.81%
	total	5,526	223 = 4.10%
Annegers *et al.*	1980	2,757	97 = 3.53%

Table 2. *Incidence of Epilepsy in Combat Injuries*

	All cases		Non-missile blast injury		Missile injury	
	Total	Fits	Total	Fits	Total	Fits
World War I						
Credner (1930)	1,990	(38%)				
Ascroft (1941)	317	107 (34%)				
World War II						
Walker Jablon (1959)	739	207 (28%)	444	107 (24%)	295	100 (34%)
Korea						
Caveness *et al.* (1962)	407	98 (24%)	196	24 (12%)	211	74 (35%)
Vietnam						
Salazar *et al.* (1985)	421	224 (53%)			421	224 (53%)

C. A. Pagni: Posttraumatic Epilepsy. Incidence and Prophylaxis

39

Table 3. *Civilian Non-missile Brain Wounds*

	Total	Fits
Evans (1962)	100	32 = 32%
Paillas *et al.* (1970)	186	59 = 32%

epilepsy within 1, 2, 3, or 4 weeks as early epilepsy[5, 12, 17, 33, 37].

Jennett[25, 26] (1969, 1975) proposed that epilepsy occurring in the first week is distinctly different from that occurring in the next weeks: fits in about 50% of the cases are focal without generalization; temporal lobe attacks were never encountered; first week epilepsy recurs in the future significantly less often than epilepsy beginning in the following weeks; the incidence of early epilepsy is higher in young children; long posttraumatic amnesia in the absence of cerebral damage increases the risk of early but not of late seizures. Thus he suggested the term "early epilepsy" should be reserved for fits within the first week. Weiss and Caveness[44] (1972) produced evidence directed at supporting that definition, and usually the term is just referred to fits beginning in the first week.

2. Early Epilepsy

2 a Incidence

The incidence of early epilepsy (in the first week or in the first two weeks) in unselected series of consecutive, usually closed, head injuries is between 3–6% (Table 4).

In severe head injuries (depressed fractures, brain contusion and lacerations, haematomas, posttraumatic unconsciousness and coma lasting at least for 48 hours) the incidence rises to a value of 8–10% (Tables 5 and 6).

In trivial injuries (i.e. injuries without initial unconsciousness or posttraumatic amnesia and not complicated by depressed fracture or intracranial haematoma – Jennett 1975) the incidence is 2–3%.

Table 5. *Incidence of Early Epilepsy in Series of Severe Head Injuries*

	Total case number	Early fits
Gros *et al.* (1955)	300	21 = 6.7%
Stowsand (1971)	1,107	107 = 9.6%
De Santis and Pagni (1976 a)	1,091	90 = 8.2%

Table 6. *Incidence of Early Epilepsy in Head Injuries of Different Severity* (from De Santis and Pagni 1976 b)

	Cases	Fits
Trivial injury	2,170	73 = 3.3%
Coma (no surgery)	249	36 = 14.4%
Surgery (laceration haematoma etc.)	127	25 = 19.6%
Total	2,546	134 = 5.3%

2 b Influence of Age on Incidence of Early Epilepsy

According to Jennett (1975) early epilepsy occurs as frequently in adults as in children under the age of 16. In my series early epilepsy occurred nearly twice as frequently under the age of 16 than in adults (Table 7).

Furthermore early epilepsy is more frequent before the age of 5 years (Table 8).

The proneness of the youngest children to seizure could explain the observation of Hendrick *et al.* (1968) and Jennett (1975) that the incidence of early epilepsy after even trivial injury is higher in younger children. On the whole however early epilepsy affects with the highest frequency children in the first three years of life (6.6%, De Santis and Pagni 1976 b).

2 c Interval from Injury to First Fit

The first early epileptic attack occurs within 24 hours of injury in about half to two thirds of the cases, if all

Table 4. *Incidence of Early Epilepsy in Series of Consecutive, Usually Closed, Head Injuries Including Children and Adults*

		Total case number	Early fits	
Hendrick *et al.* (1968)	(child)	4,195	252	6.00%
Courjon (1970)		Thousands	?	4.00%
Jennett (1975)		1,000	46	4.60%
De Santis – Pagni (1976 b)	(child)	2,546 } 5,526	218	3.94%
De Santis *et al.* (1979 a)	(adult)	2,980 }		
Annegers *et al.* (1980)		2,747	58	2.10%

Table 7. *Influence of Age on Incidence of Early Epilepsy*

	Jennett (1975)			De Santis Pagni (1976 b) De Santis *et al.* (1979 a)			Annegers *et al.* (1980)		
< 16 years	202	11	5.4%	2,546	134	5.3%	1,132	29	2.6%
> 16 years	784	35	4.5%	2,980	84	2.8%	1,615	29	1.8%
Total	986	46	4.6%	5,526	218	3.9%	2,747	58	2.1%

Table 8. *Influence of Age on Incidence of Early Fits in Children*

	Jennett (1975)		De Santis and Pagni (1976 b)	
	Cases	Fits	Cases	Fits
< 5 years	75	7 = 9%	1,001	75 = 7.5%
6–16 years	127	4 = 3%	1,149	59 = 5.1%

Table 9. *Time of Onset of Early Epilepsy at Different Ages* (Jennett 1975)

Jennett	(1975)	1 hour	1–24 hours	> 24 hours
All ages	407	115 = 28%	136 = 33%	156 = 38%
< 5 years	78	24 = 31%	36 = 46%	18 = 23%
6–15 years	101	41 = 40%	38 = 38%	22 = 22%
> 16 years	228	50 = 22%	62 = 27%	116 = 51%

Table 10. *Time of First Fit After Injury in Children up to 15 Years* (from De Santis and Pagni 1976 b)

First hour	33 = 24.6%	} 79 = 58.9%	}
2–6 hours	46 = 34.3%		} 96 = 71.7%
7–12 hours	9 = 6.7%	} 17 = 12.7%	
13–24 hours	8 = 5.9%		
over 24 hours	38 = 28.3%		38 = 28.3%

series that include children and adults are considered (Table 9).

In children epilepsy begins within the first 24 hours of injury more often than in adults and also more often during the first 6 hours after injury (Table 10).

Broadly speaking early fits occur more often in the first hour(s) after injury in depressed fractures and in patients with marked loss of consciousness lasting more than 24 hours[26] or in coma but not harbouring intracranial haematoma[11]; occur beyond the first day if there is an intracerebral haematoma or a haemorrhagic brain laceration and very often after operation[11].

2 d Severity of Trauma and Occurrence of Early Seizures

In the past many efforts have been made to identify if there was any feature of trauma and its consequences that related to an increase of occurrence of early fits. A complete and superb analysis of injury features influencing the occurrence of early seizure was made by Jennett (1975) before the introduction of the CTScan. Broadly speaking injuries which are more severe or complicated (loss of consciousness, linear or depressed fracture, dural tear, brain contusion and haematoma, neurological signs) are associated with an increased risk of epilepsy (Table 6). Some of the relevant conclusions are:

a) Early epilepsy rarely follows injuries without any period of unconsciousness or with short posttraumatic unconsciousness except in children under the age of 5 and injury with a depressed fracture[26, 27];

b) Loss of consciousness of more than 24 hours increases the early epilepsy rate from 2–6% to 11–14%. If there is no loss of consciousness or concussion is very mild, early epilepsy is more frequent in children under 5 years than in adults[1, 11, 12, 26];

c) Linear fracture increases the early epilepsy rate from 2% to about 7%, while the site of linear fracture does not affect the incidence of early epilepsy rate[26];

d) Depressed fracture increases further the occurrence of early epilepsy up to 10–11%[26];

e) Intracranial haematomas and haemorrhagic contusion lacerations increase the early epilepsy rate up to 20–30% both in children and in adults[1, 12, 26];

f) Incidence of early epilepsy is significantly higher after subdural and intracerebral haematoma or haemorrhagic laceration than after extradural haematoma reaching in certain series the value of 45%.

g) Neurological signs, mainly the focal ones, increase the incidence of early fits regardless of age[26].

As far as interaction of the various features of injury on seizure occurrence, Jennett (1975) concluded that the early epilepsy rate after depressed fracture and/or intracranial haematoma or brain haemorrhagic laceration is not affected by any other feature of the injury (age of the patient, duration of unconsciousness, presence of focal signs, whether the fracture was closed or compound and whether the dura was torn or not).

However studies on traumatic epilepsy made in the CTScan era have demonstrated that the main factor for development of early and late posttraumatic epilepsy is focal haemorrhagic brain damage[10]. On the other hand prolonged posttraumatic unconsciousness should not be considered "per se" a risk factor for epilepsy if not associated with focal brain damage[15].

2e Type of Early Fits

Fits are of focal type in 60–80% of the cases[12, 16, 17, 26, 36, 41]. Grand Mal and tonic generalized seizures are observed in 20–40% of the cases. Focal attacks are more frequent in missile injuries (up to 80% of the cases) than in non missile. However focal attacks, including alternating bilateral tonic-clonic seizures are more frequent in children than in adults[12]. Psychomotor or Petit Mal attacks were never observed.

2f Significance of Early Seizures as far as Outcome is Concerned

Two kinds of attacks must be distinguished:
i) sporadic fits (single or repeated spaced fits)
ii) recurring fits and status epilepticus

Table 12. *Outcome of Traumatic Coma with Sporadic Fits or Status* (De Santis and Pagni 1976 a)

Traumatic coma with early seizures: 90 cases

	Cases	Deaths
Sporadic fits (single or repeated)	54	21 = 39%
Status or recurring fits	36	18 = 50%

Table 13. *Incidence of Sporadic Seizures and Status in Children and Adults*

	De Santis Pagni (1976a) – Child 134/2170	De Santis *et al.* (1979a) – Adult 84/2980
Single or repeated fits	101 = 75%	74 = 87%
Recurring fits or status	33 = 25%	10 = 13%
	134 = 100%	84 = 100%

2fi Sporadic fits: In a small number of cases and usually in children under 5, epilepsy may follow a trivial injury[26]. Usually early epilepsy is associated with brain contusion or laceration and with intracranial haematoma. Occurrence of an early fit is not "per se" a reason for suspecting an intracranial haematoma. An early fit is not usually the only or the first sign of an intracranial haematoma: other signs of the complication are associated with the fit. Furthermore early epilepsy in itself does not make the prognosis worse as far as mortality is concerned. In my series of traumatic coma operated upon for intracranial haematomas the death rate was superimposable in cases with or without early sporadic seizures (Table 11).

Table 11. *Death Rate in Operated and Not Operated Traumatic Comas With and Without Early Seizures*[11]

1,091 Traumatic Comas	No seizure 1,001 cases		Early seizure 90 cases	
	Cases	Death	Cases	Death
Operated on	614	342 = 55%	48	27 = 56%
Not operated on	387	185 = 47%	42	12 = 29%
Total	1,001	527 = 53%	90	39 = 43%

On the contrary Jennett (1975) stated there was in his series an increased mortality rate associated with early epilepsy. That might be due almost wholly to the association between acute intracranial haematoma and early epilepsy in a general series not including only patients in coma but also trivial injuries. In fact in my series the lowest mortality rate was observed in patients with early fits in coma not submitted to surgery (Table 11).

2 f ii Recurring seizures and epileptic status: Epileptic status, whatever the origin and especially in children, carries a risk of persistent brain damage if active treatment is not instituted promptly. Mortality in the series of Jennett (1975) was not higher in patients with status than in other types of early seizures. In my series on the contrary mortality was higher in patients with status or affected by fits recurring at short intervals. Most of those patients had been operated upon for brain lacerations, intracerebral or subdural haematoma; status and recurring seizures developed postoperatively (Table 12).

Epileptic status is more frequent in children (Table 13), especially under 5, than in adults; furthermore, while in adults the status is usually associated with severe cerebral damage[12] in children under 5 status and recurring seizures — both as generalized or as focal fits — have been observed also after mild head injury.

2 g Early Seizures and Predisposition to Late Epilepsy

For a long time early fits were considered of no relevance as far as late epilepsy is concerned. But it is now clear late epilepsy is significantly more common in patients who had suffered early fits (Table 14), both in children and in adults. Jennett (1975) believes early epilepsy carries a significantly increased risk of late epilepsy in adults but not in children.

The influence of early epilepsy on the incidence of late epilepsy varies greatly according to the different features of the injury and the type of early fits. The relevant data may be summarized in the following statements:

i) As far as any feature of the injury is concerned:
i 1. Incidence of late epilepsy after prolonged unconsciousness and coma, focal signs, intracranial haematoma, brain laceration, and missile injuries is nearly equal both in patients who have and do not have early epilepsy. In other words such types of injuries carry a high risk of late seizure whether or not early traumatic epilepsy occurs[14–26].
i 2. In depressed fractures, early epilepsy increases significantly the risk of late epilepsy[26].
i 3. If there had been an early fit the risk of late epilepsy is increased greatly in trivial and not complicated injuries which would otherwise have a low risk of late epilepsy. That applies both to children, even under 5, and to adults[14, 14 bis, 26].

ii) As far the type of early fit is concerned:
Opinions on the influence of the type of early fit on the risk of late epilepsy diverge. According to Jennett (1975) a single fit is equally likely to be followed by late epilepsy as repeated fits or even status epilepticus. In my series of children on the contrary occurrence of late epilepsy was significantly higher after repeated fits or status (Table 15).

Evans (1962) reported that fits in the first 24 hours

Table 14. *Late Epilepsy After Early Fits*

	Early epilepsy	No early epilepsy
Children		
Hendrick Harris (1968)	37/124 = 21%	2%
Stowsand (1971)	8/40 = 20%	
Jennett (1975)	20/118 = 17%	8/230 = 4%
De Santis *et al.* (1979 b)	13/64 = 20%	
Adults		
Evans (1963)	5/14 = 36%	7/134 = 5%
Weiss Caveness (1972)	8/26 = 31%	
Jennett (1975)	39/120 = 33%	22/663 = 3%
De Santis *et al.* (1983)		21/472 = 4%

Table 15. *Number of Early Fits and Status in Relation to Late Epilepsy*

	One fit	Repeated fits	Status
Jennett (1975)	24/97 25%	31/121 25%	5/21 24%
De Santis *et al.* (1979 b)	1/24 4%	5/19 26%	7/18 38%

Table 16. *Time of First Early Fit and Occurrence of Late Epilepsy*

	Interval to early epilepsy	
	< 24 hours	> 24 hours
Jennett (1975)	51/251 20%	39/156 25%
De Santis *et al.* (1979 b) (Children)	7/44 15%	6/18 33%

C. A. Pagni: Posttraumatic Epilepsy. Incidence and Prophylaxis

43

after injury are usually not followed by late seizures. Jennett (1975) states that the risk of epilepsy is increased by early fits occurring in the first 24 hours, but that time of early fits (in the first hour, day or week) does not affect the incidence of late epilepsy. In my series in children the risk of late epilepsy was increased greatly by early fits occurring after 24 hours from injury[12] (Table 16). As far as so-called immediate fits are concerned (that is generalized fits occurring within moments of injury following mild injury, generally in adults) they are not usually followed by late epilepsy[26].

3. Late Epilepsy

3 a Incidence of Late Epilepsy

Estimates of the incidence of epilepsy after closed, civilian or non-missile injuries vary widely. For unselected series of non-combat head injuries admitted to hospitals it varies between 2 and 23% of the cases (Table 17). Insurance series report very low figures of 0.1% of posttraumatic epilepsy. Incidence is higher in penetrating missile injuries (up to 53%, Salazar et al. 1985).

Hence large differences are due mainly to different criteria for recruitment of patients: inclusion or exclusion of cases with early seizures or of cases presenting even a unique fit in the weeks following injury or only of cases with repeated attacks; various length of follow-up; study of neurosurgical departments series versus series including mainly outpatients; inclusion of high number of open injuries. However with growing knowledge it became evident that the incidence of late posttraumatic epilepsy varies widely according to the type of injury and efforts were made to identify risk factors.

3 b Risk Factors

Before the introduction of CTScan, risk factors that increased the incidence of late epilepsy were considered to be acute intracranial haematomas, early epilepsy, depressed fracture, duration of loss of consciousness or coma, focal signs, linear fracture, penetrating injury, location of the cerebral lesion and torn dura. The in-

Table 17. Late Epilepsy After Civilian Injuries

Penfield and Shaver (1945)	/407	2.4%
Jennett (1975)		5.0%
Annegers et al. (1980)	51/2,747	1.85%
De Santis et al. (1983)	21/472	4.44%
Paillas et al. (1970)	188/2,145	9.0%
Guidice and Berchou (1987)	38/164	23.0%

Table 18. Risk Factors for Late Posttraumatic Epilepsy

	Incidence of late seizures (%)
Penetrating injury caused by missile	53
Intracerebral haematoma – laceration	39
Focal brain damage on early CTScan	32
Early seizures	25
Depressed fracture – torn dura	25
Extradural or subdural haemorrhage	20
Focal signs (hemiplegia, aphasia, ...)	20
Depressed skull fracture	15
Loss of consciousness > 24 hours	5
Linear fracture	5
Mild concussion	1

troduction of CTScan examination gave another clue to the analysis of the risk factors. It was demonstrated that duration of unconsciousness or coma is not per se a relevant risk factor for epilepsy provided cerebral contusion is absent[15] while focal parenchymal damage on an early CTScan after injury increases significantly, as is shown by statistical analysis, the risk of posttraumatic epilepsy[10]. In the series of De Santis et al. (1988) the CTScan was either normal or showed diffuse brain injury or brain swelling without focal brain damage: no patient presented with late seizure at a mean of 4.1 years follow-up. No report is available in the literature on the relationships between duration of coma due to extracerebral haematoma (extradural haematoma, subdural haematoma) with hippocampal hernia and posterior cerebral artery territory ischemia and incidence of epilepsy. In the series of d'Alessandro et al. (1982) out of 31 patients presenting with CTScan focal brain damage (hypodense contusion or haemorrhagic contusion), 10 (32%) developed posttraumatic epilepsy. Even the total brain volume loss on CTScan (in penetrating injuries – Salazar et al. 1985) has a significant correlation with seizures. A total lost volume of less than 25 cc was followed by seizures in 42% of patients, while 80% of patients with a loss of more than 75 cc had late epilepsy[38].

On the basis of the analysis of the pathological characteristics of injury, CTScan data and neurological outcomes, predictive formulae have been developed to anticipate occurrence of late epilepsy in any given patient[45]. In Table 18 have been collected some risk factors for late epilepsy which have been identified[14, 26, 38, 45] both in closed and in open head injuries.

The risk of late seizures increases if risk-factors are associated and if cerebral damage involves certain brain regions. For instance penetrating missile injuries damaging the centro-parietal region gave rise to late epilepsy in more than 75% of patients[38].

The association between depressed fracture, early seizures and focal signs increases greatly the risk of late epilepsy. On the contrary, strangely enough, retained bone fragments (in penetrating missile injuries), use of a dural graft, cranioplasty, brain abscess, or family history of epilepsy seem to have no impact on incidence of late epilepsy[38]. I did not find any report analyzing the influence of interval between injury and operation for depressed fracture, open head trauma, penetrating injury, haematoma on the incidence late epilepsy.

3 c Time of Onset of Late Seizures

Many studies have shown that over 50% of late seizures occur within one year of injury, 70–80% within two years. Thereafter for about 10 years 3–5% new seizures occur yearly. Thus in over 15% of patients the first seizure may occur more than 5 years after injury; and another 5% will not manifest epilepsy until 10 or more years later. That seems to be valid both for closed and penetrating missile injuries[26, 38]. Annegers *et al.* (1980) compared the number of cases of late traumatic epilepsy in their series (51 cases out 2,747) with the number of new cases of seizures that would be expected in the general population of the same age and sex composition and during the time equivalent to their follow-up. These calculations showed that there was a 3.6 fold increase of risk after head injury; but the risk, which was still 4.4 at five years of injury, decreased with time. After 5 years the risk was not significantly greater than normal in their series of closed head trauma with moderate to mild head injuries; the number of patients with severe head trauma followed for more than 5 years was too small for evaluation of the risk level, which on the contrary seems to be still raised for penetrating injuries and severe brain damage[38].

4. Number of Cases of Posttraumatic Epilepsy in General Series of Epileptic Patients

In the last 30 years many papers on the aetiology of epilepsy have been published. A summary of actual knowledge may be found in the paper of Bergamini *et al.* (1977). Aetiological factors are recognizable only in about 35–45% of the cases (Table 19).

In spite of some differences due to inclusion or exclusion of epilepsy due to tumors, vascular lesions,

Table 19. *Epileptic Population*

		Gudmundsson (1966)	Bergamini *et al.* (1976)
Etiology	known	339 = 34%	782 = 44%
	unknown	648 = 66%	1,003 = 56%
	total	987 = 100%	1,785 = 100%

Table 20. *Aetiologic Factors*

	Gudmundsson (1966)	Bergamini *et al.* (1977)
Birth injury	142 = 38%	318 = 41%
Head injury	82 = 23%	218 = 28%
Infectious disease	52 = 14%	213 = 27%
CNS malformat. and disease	36 = 9%	33 = 4%
Tumours	20 = 5%	not included
Vascular lesions	38 = 10%	not included
Total	370 = 100%	782 = 100%

metabolic diseases and so on, posttraumatic epilepsy accounts for about 25% of the cases for which an aetiological factor is identifiable (Table 20).

Thus broadly speaking a head injury sustained after birth is the cause of seizures in 10–12% of the overall epileptic population.

5. Type of Fits in Late Posttraumatic Epilepsy

It is usually difficult to collect data from the literature in respect of the type of posttraumatic fits. Generally data on the clinical characteristics of the attacks are lacking. In Table 21 some available data from the literature obtained in series analyzed by retrospective anamnestic scrutiny and prospective study are collected.

In all the reported series, generalized seizures (including cases of focal attacks becoming secondarily generalized) account for 60–70% of the attacks and partial seizures (with elementary or complex symptomatology) account for 30–40% of the cases.

6. Prophylaxis of Posttraumatic Epilepsy

Experimental researches showed that epilepsy due to kindling, cobalt or alumina cream foci may be prevented by phenytoin, phenobarbital or carbamazepine if drug administration begins immediately after the ap-

Table 21. *Type of Fits in Late Posttraumatic Epilepsy*

	Retrospective Bergamini et al. (1977)	Prospective Courjon (1970)	Jennett (1975)	Salazar et al. (1983)
GM	142 = 65%	32 = 40%	289 = 60%	68 = 30%
F-GM	? = ?	24 = 30%	? = ?	89 = 40%
PM	0 = 0	0 = 0	0 = 0	0 = 0
HHE	3 = 1%	0 = 0	0 = 0	0 = 0
PaE	43 = 23%		102 = 21%	22 = 10%
PaC	30 = 14%	24 = 30%	90 = 19%	44 = 20%
Total	218 = 100%	80 = 100%	481 = 100%	223 = 100%

GM: generalized seizures. F-GM: focal at onset becoming generalized. PM: petit mal and Lennox-Gastaut syndrome. HHE: hemigrand mal. PaE: partial seizures with elementary symptomatology. PaC: partial seizures with complex symptomatology.

plication of the epileptogenic agent to block the "ripening" of the epileptic focus[3, 35]. Early studies in the human seemed to show a significant benefit from the use of phenytoin or phenobarbital and phenytoin[47]. Servit and Musil (1981) reported that low-dose phenytoin and phenobarbital treatment for 2 years has a significant effect in preventing seizures even years after the drug has been discontinued. As a result of these studies the Czechoslovakian government mandated prophylactic antiepileptic drugs in severe head injuries. Other trials showed[29] that in adults with serum concentration of phenytoin or phenobarbital in the therapeutic range for at least 6 months "there was a trend toward a decreased seizures frequency ... during an 18-months follow-up period after the drug has been discontinued ...". Many criticisms were raised against these studies: no double-blind studies; no randomized assignment to treatment and to non-treatment groups; drug's serum concentrations were not always monitored; much data was collected retrospectively. In 1983 Young et al. made the first randomized double-blind placebo-controlled study in patients treated by phenytoin as soon as possible after injury and within 24 hours of hospital admission. The conclusion was that phenytoin does not significantly decrease the incidence of posttraumatic epilepsy in children. These experiences prompted scepticism of the prophylactic effect of phenytoin and phenobarbital. The conclusion was reached that prophylactic drug treatment is not useful: most of the patients were being exposed unnecessarily (toxicity; cost) to antiepileptic medication. Patients with mild trauma should not be treated. Only patients with a seizure within two years of injury should be considered

at high risk for recurrence of seizures and placed on antiepileptic drugs. Patients with a later onset of seizures should not be treated until a second seizure occurs[29].

Intracortical injection of blood or blood derivatives provokes cortical epileptic foci and recurring seizures in the cat [21, 28, 46]. The histological picture of the lesion is similar to that found in human posttraumatic epileptogenic foci[31]. Thus the iron-induced epilepsy was considered a model of posttraumatic epilepsy[46]. Even the development of an alumina cream focus is associated with microhaemorrhages and hemosiderin deposition[22]. Iron epileptogenesis might be due to peroxidation of microsomal membrane proteinlipid complexes[42]. Bergamini and Mutani (1985) showed that the administration of valproate slows the "ripening" of the epileptic focus and reduces dramatically the incidence and frequency of seizures both in alumina cream and cobalt epilepsy. Price (1980) attempted prophylaxis by valproate. His results are not reported in any paper on prophylaxis of posttraumatic epilepsy. In a consecutive series of 143 patients with high risk of epilepsy (15–55%) sodium valproate 600–1,500 mgrs a day was given to the patients. During a two years follow-up no patient developed epilepsy in spite of predicted 26% risk. In 8% of the patients carbamazepine was substituted for valproate because of toxic signs. But recently Salazar (1989) reported valproate is of no effect in preventing posttraumatic epilepsy. However he stated that intravenous administration of superoxide dismutase (an antioxidant agent) in animals blocks the ripening of epileptic foci. Clinical experience is still lacking. We are still compelled to agree for the moment with Caveness' (1976) statement: "the only certain way to prevent traumatic epilepsy is to reduce head injuries".

References

1. Annegers JF, Grabow JD, Groover RV, Laws ER Jr, Elveback LR, Kurland LT (1980) Seizures after head trauma: a population study. Neurology 30: 683–689
2. Ascroft PB (1941) Traumatic epilepsy after gunshot wounds of the head. Br Med J 1: 739–744
3. Bergamini L, Mutani R (1985) Epilepsie. In: Bonavita V, Quattrone A (eds) Terapia Medica delle Malattie del Sistema Nervoso. Piccin Editore, Padova, pp 360–392
4. Bergamini L, Bergamasco B, Benna P, Gilli M (1977) Acquired etiological factors in 1785 epileptic subjects: clinical-anamnestic research. Epilepsia 18: 437–444
5. Caveness WF (1963) Onset and cessation of fits following craniocerebral trauma. J Neurosurg 20: 570–583
6. Caveness WF (1976) Epilepsy a product of trauma in our time. Epilepsia 17: 207–215
7. Caveness WF, Walker AE, Ascroft PB (1962) Incidence of posttraumatic epilepsy in Korean veterans as compared with those from World War I and World War II. J Neurosurg 19: 122–129
8. Courjon J (1970) A longitudinal electro-clinical study of 80 cases of posttraumatic epilepsy observed from the time of the original trauma. Epilepsia 11: 29–36
9. Credner L (1930) Klinische und soziale Auswirkungen von Hirnschädigungen. Z Ges Neurol Psychiat 126: 721–757
10. d'Alessandro R, Tinuper P, Ferrara R, Cortelli P, Pazzaglia P, Sabattini L, Frank G, Lugaresi E (1982) CTScan prediction of late posttraumatic epilepsy. J Neurol Neurosurg Psychiatry 45: 1153–1155
11. De Santis A, Pagni CA (1976 a) Valore prognostico delle crisi convulsive precoci nel coma traumatico. Riv Neurol 46: 400–406
12. De Santis A, Pagni CA (1976 b) Osservazioni su 134 casi di traumatizzati cranici in eta' infantile con crisi convulsive precoci. Atti del VII Congresso Naz. della Soc. Ital Neuropsich Infantile Ediz Centro Min Medica, Torino, pp 269–273
13. De Santis A, Cappricci E, Granata G (1979 a) Early posttraumatic seizures in adults. Study of 84 cases. J Neurosurg Sci 23: 207–210
14. De Santis A, Pagni CA, Rampini P (1979 b) Osservazioni sul controllo a distanza di 64 pazienti con crisi precoci posttraumatiche. Riv ital Elettroencef Neurofisiol Clin 2: 43–46
14 bis. De Santis A, Rampini P, Granata G, Sina C, Biasini A, Ravagnati L (1983) Epilessia posttraumatica in eta' adulta: Controllo su una serie di traumatizzati cranici in 6 anni consecutivi. Riv ital Elettroencef Neurofisiol Clin [Suppl] 1: 287–288
15. De Santis A, Rampini P, Sganzerla EP (1988) Prolonged posttraumatic unconsciousness, diffuse brain injury and epilepsy. Boll Lega ital Epilessia 62–63: 79–82
16. Evans JH (1962) Posttraumatic epilepsy. Neurology (Minneap) 12: 665–674
17. Evans JH (1963) The significance of early posttraumatic epilepsy. Neurology (Minneap) 13: 207–212
18. Gros CI, Passouant P, Cadilhac J, Vlahovitch B, Roilgen A (1955) Epilepsie precoce posttraumatique. Neuro-chirurgie 1: 325–326
19. Gudmundsson G (1966) Epilepsy in Iceland. A clinical and epidemiological investigation. Acta Neurol Scand [Suppl 25] 43: 6–124
20. Guidice MA, Berchou RC (1987) Posttraumatic epilepsy following head injury. Brain Injury 1: 61–64
21. Hammond EJ, Ramsay RE, Villareal HJ, Wilder BJ (1980) Effects of intracortical injection of blood and blood components on the electrocorticogram. Epilepsia 21: 3–14
22. Harris AB (1972) Degeneration in experimental epileptic foci. Arch Neurol 26: 434–449
23. Hendrick EB, Harris L (1968) Posttraumatic epilepsy in children. J Trauma 8: 547–558
24. Hendrick EB, Harwood-Hash DCF, Hudson AR (1968) Head injuries in children: A survey of 4465 consecutive cases at the Hospital for Sick Children, Toronto, Canada. Clin Neurosurg 15: 46–65
25. Jennett WB (1969) Early traumatic epilepsy. Lancet i: 1023–1026
26. Jennett B (1975) Epilepsy after non-missile head injury. Heinemann Publ, London, 179 p
27. Kollevold T (1976) Immediate and early cerebral seizures after head injuries. J Oslo City Hosp 26: 99–114
28. Lange SC, Neafsey EJ, Wyler AL (1980) Neuronal activity in chronic ferric chloride epileptic foci in cats and monkey. Epilepsia 21: 251–259
29. Oles KS, Penry JK (1985) Pharmacological prophylaxis of posttraumatic seizures. In: Johnson RT (ed) Current therapy in neurologic disease 1985–1986. Decker and Mosby, Philadelphia, pp 46–51
30. Paillas JE, Paillas N, Bureau M (1970) Posttraumatic epilepsy. Introduction and clinical observations. Epilepsia 11: 5–16
31. Payan H, Toga M, Berard-Badier M (1970) The pathology of posttraumatic epilepsies. Epilepsia 11: 81–94
32. Penfield and Shaver (1945) Quoted in Jennett 1975
33. Phillips G (1954) Traumatic epilepsy after closed head injury. J Neurol Neurosurg Psychiatry 17: 1–10
34. Price DJ (1980) The efficiency of sodium valproate as the only anticonvulsant administered to neurosurgical patients. In: Parsonage MJ, Caldwell ADS (eds) The place of sodium valproate in the treatment of epilepsy. Academic Press, London, pp 23–34
35. Rapport RL, Ojemann GA (1975) Prophylactically administered phenytoin. Effects on the development of chronic cobalt-induced epilepsy in the cat. Arch Neurol 32: 539–548
36. Rish BL, Caveness WF (1973) Relation of prophylactic medication to the occurrence of early seizures following craniocerebral trauma. J Neurosurg 38: 155–158
37. Russell WR, Whitty CWM (1952) Studies in traumatic epilepsy. Part I: Facts influencing the incidence of epilepsy after brain wounds. J Neurol Neurosurg Psychiatry 15: 93–107
38. Salazar AM, Jabbari B, Vance SC, Grafman J, Amin D, Dillon JD (1985) Epilepsy after penetrating head injury. I. Clinical correlates: A report of the Vietnam Head Injury Study. Neurology 35: 1406–1414
39. Salazar AM (1989) Oral presentation at the "Symposium on posttraumatic epilepsy", Pisa, May 17
40. Servit Z, Musil F (1981) Prophylactic treatment of posttraumatic epilepsy: results of a long term follow-up in Czechoslovakia. Epilepsia 22: 315–321
41. Stowsand D (1971) Paresen und epileptische Reaktionen im Initialstadium des Hirn-Traumas. G Thieme, Stuttgart

42. Victoria EJ, Barber AA (1969) Peroxidation of microsomal membrane proteinlipid complexes. Lipids 4: 389

43. Walker AE, Jablon S (1971) A follow-up study of head wounds in World War II. Government printing office

44. Weiss GH, Caveness WF (1972) Prognostic factors in the persistence of posttraumatic epilepsy. J Neurosurg 37: 164–169

45. Weiss GH, Feeney DM, Caveness WF, Dillon D, Kistler JP, Mohr JP, Rish BL (1983) Prognostic factors for the occurrence of posttraumatic epilepsy. Arch Neurol 40: 7–10

46. Willmore LJ, Sypert GW, Munson JB (1978) Recurrent seizures induced by cortical iron injection: a model of posttraumatic epilepsy. Ann Neurol 4: 329–336

47. Young B, Rapp R, Brooks WH, Madauss W, Norton JA (1979) Posttraumatic epilepsy prophylaxis. Epilepsia 20: 671–681

48. Young B, Rapp RP, Norton JA, Haack D, Walsh JW (1983) Failure of prophylactically administered phenytoin to prevent late posttraumatic seizures. J Neurosurg 58: 236–241

Correspondence: Prof. Dr. Carlo A. Pagni, 2nd Chair of Neurosurgery, University of Torino, Via Zuretti 29, I-10126 Torino, Italy.

Acta Neurochirurgica, Suppl. 50, 48–51 (1990)

Controversies in Posttraumatic Epilepsy*

A. Martins da Silva[1], A. Rocha Vaz[2], I. Ribeiro[2], A. R. Melo[2], B. Nune[3], and M. Correia[3]

[1] Servico de Neurofisiologia, Hospital de Santo Antonio (Porto) and Sector Fisiologia Humana, Instituto Ciencias Biomedicas Abel Salazar, Universidade do Porto, Porto, Portugal
[2] Servico de Neurocirurgia and [3] Servico de Neurologia, Hospital de Santo Antonio, Porto, Portugal

Summary

In civilian accidents the factors involved in the origin of posttraumatic epilepsy are controversial. In this study of 506 consecutive head trauma patients and 101 epileptic patients developing seizures after head trauma, we have examined the importance of the duration of coma, type of seizure and drug therapy, time to first seizure, age and focal lesions and compared our results with the literature. The importance is stressed of focal lesions and of neurological deficits for the origin of posttraumatic seizures.

Keywords: Epilepsy; head-trauma; posttraumatic epilepsy; risk factors.

Introduction

Head trauma has been investigated systematically as the cause of epilepsy in veterans who survived world wars and other conflicts including Vietnam[2, 10–13]. These authors suggested that this type of epilepsy was more frequent with penetrating head injuries[12], with injuries accompanied by prolonged loss of consciousness[12] or with loss of cerebral tissue[10]. Data from civilian accidents (for example road traffic or work accidents) further complicate the definition of risk factors, but at least the series are consistent in their reported frequency and prevalence of epilepsy caused by head trauma. In these series the role of a dural tear, posttraumatic amnesia and of depressed fractures have been highlighted as important factors[6]. In both populations the epileptic attacks are considered as acute or early seizures, if they appear on the first day or, with some authors, during the first week following trauma[6], and late seizures if they appear subsequently.

An overall incidence of 7–10% for posttraumatic epilepsy is accepted for civilian accidents[5]. In both populations (civilian and war) the incidence of acute sei-zures is lower than late seizures[2, 6]. This value is about 7 times greater in penetrating head injuries[10]. The frequency of acute seizures was reported to be higher either in all children[9] or only in school children[1]. These age differences were not seen by others[6, 8].

It is not only in children that some divergencies are evident in posttraumatic epilepsy studies. The relative importance of various risk factors varies. In some studies, namely from Majkowski, the role of familial susceptibility was considered as predictive for post-traumatic epilepsy[9]. The influence of some risk factors has been studied to determine the probabilities of a patient developing posttraumatic epilepsy. Feeney and Walker developed a mathematical approach to calculate the probabilities of a patient developing seizures after having had one head trauma[4]. From data already published they established equations based on a constant probability model. This model was later used by Weiss et al.[13], who found partial agreement between the real and the mathematical expected values.

The literature on post-traumatic epilepsy still leaves the relative importance of certain risk factors for post-traumatic epilepsy controversial. The following areas merit further investigation:

i) The role of coma (severity and duration);

ii) The role of acute seizures on the appearance of late posttraumatic epilepsy;

iii) The changes in seizure type (generalized after trauma and focal later on);

iv) The greater incidence in children;

v) The role of focal haemorrhagic lesions and of the blood brain barrier;

vi) The value of prophylactic treatment for the prevention of epilepsy.

We wish to present our preliminary results that shed

* This work was supported by JNICT (Portuguese Scientific and Technological Committee) grant 87183.

some light on these controversial questions except for the last two. The value of prophylactic treatment will be assessed when sufficient numbers of prospective patients are available. The role of blood brain barrier and of focal haemorrhagic lesions is being investigated prospectively by clinical, neuroradiological and neurophysiological approaches, as well as experimentally. These studies are not included here.

Materials and Methods

Two groups of patients were investigated. The first one included 506 consecutive patients (418 males; 88 females) who were admitted to the head trauma intensive care unit. In this group we studied the incidence of posttraumatic epilepsy. Only 15% of them were children, 25% adolescents and 70% were less then 50 years old. The cause of the trauma was, in 60% of the cases road traffic accidents and in 20% work or domestic accidents. 35% of the patients had multiple trauma. They have been followed up for more than 3 years after the trauma in the outpatient clinic. In this group we determined the type and severity of head trauma defined according to the level of consciousness (scored by means of the Glasgow Coma Scale – GCS – criteria) and the presence of neurological deficits. In patients having had seizures we studied the possible factors involved, the time interval between trauma and seizures the type and the modifications of seizures if any, the medication and the follow-up.

The second group comprised data from 101 adults followed in the outpatient clinic for epilepsy and having had head trauma (caused mainly by road traffic accidents) and complaining of seizures thereafter. In this group, we analysed the type and severity of the trauma also defined by the level of consciousness (GCS), neurological deficits, the interval between the accident and the date of seizures beginning, the type of seizures, their modification if any, and their evolution.

Results

In the first group analysed (506 patients) the level of consciousness defined by the Glasgow Coma Scale was lower than 9 points in 28% of the cases and in 51% of them was equal or higher than 13 points. From the whole group, 50% of the patients have had localised cerebral lesions: closed focal contusions – 24%; depressed fractures – 14%; penetrating head injury – 12%. The severity of trauma and the type of lesions are summarised in Table 1.

62 patients, representing 12% of the total, had epileptic seizures. Almost one third of them (20 patients) had seizures in the first 24 hours after trauma (acute seizures), and 42 patients began seizures more than 24 hours after (later seizures). Half of the patients having had acute seizures had late seizures. In the whole group of patients with seizures, 20% started the attacks more than one year after the trauma and 5% more than two years later. The population with seizures is distributed as follows: 10% were children, 20% adolescents and 35% aged between 19–49 years old. In each age group, seizure incidence was determined by dividing the number of patients with seizures by the total number of patients in the same group. For children, we found that this incidence was 8%. In the group of acute seizures 9 out of 20 patients (45%) had focal lesions (penetrating head injury, focal contusion, depressed fracture) and 9 patients (45%) had neurological deficits. The group of patients with seizures included 10 with a GCS score lower than 9 and 7 higher than 13 points. In the patients having had acute seizures their attacks were generalized in 13 cases (65%). In the group of patients having chronic seizures the figures are different: 47% of the patients (20 out of 42) had focal lesions and 65% (27 out of 42) had neurological deficits. In this group 18 patients had GCS score lower than 9 points and 17 higher than 13. The seizures, in this group, began or remained partial in 32 patients (76% of the cases) and generalized in 10 (24%). Medication was given in 50% of the cases with acute seizures. In total, 80% of patients have been treated. In Table 2 the characteristics of the seizures, type and age distribution, are summarised.

The second group analysed included 101 epileptic patients referred to the outpatient clinic with a history of seizures after head trauma. In this group, the patients having chronic seizures were followed up for more than 5 years. The severity of the trauma, according to the

Table 1. *Severity of Trauma and Type of Lesion in Consecutive Studies* (% of 506 Patients)

Severity (GCS)		Lesion	
<9	28%	Focal contusion	24%
9–12	21%	penetr. head inj.	12%
>13	50%	Depres. fracture	14%
		Extradural haemat.	7%
		Subdural haemat.	5%
		Diffuse contusion	33%
		Others	5%

GCS = Glasgow Coma Scale.

Table 2. *Seizure Type and Age Distribution in Consecutive Studies* (62 Patients)

Type		Age groups	
Acute	20 (32%)	Children	10%
(50% had late seizures)		Adolescents	20%
		Adults (19–49)	35%
Late	42 (68%)	Adults (>49)	35%

Table 3. *Severity of Trauma* (No. Patients = 101)

	GCS	Neurological deficit		Cerebral lesions	
	n = 80				
< 9	32%	Yes	70%	Focal lesions	56%
9–12	18%	No	30%	General contusions	12%
> 12	50%			Multiple	32%

Table 4. *Time Interval Between Trauma and Seizures*

Acute (< 24 h)	15%
(60% had late seizures subsequently)	
24 h–1 week	12%
1 week–1 month	5%
1 month–6 months	13%
6 months–1 year	14%
1 year–2 years	10%
2–5 years	12%
> 5 years	17%
Not well defined	2%

GCS, was well defined in 80 patients; it was higher than 12 points in 50% of the patients and lower than 9 in 32%. Cerebral focal or multi-lobar lesions (contusions, intracerebral haematoma, penetrating head injury, complicated depressed fracture) were present in 56% of the cases. History of neurological deficits was inferred or still present in 70% of the patients. The summary of these findings is shown in Table 3.

The seizures began within the first 24 hours after the trauma in 15% of the patients. Of those, 60% have had further seizures subsequently. In 12% of the patients the seizures appeared between the second day and the end of the first week after the trauma; in 5% the seizures began between one week and one month; in 13% between one and six months; in 14% between six months and one year; in 10% between one and two years; in 12% between two and five years; and in 17% more than 5 years after the trauma. In only 2% of the patients was this definition not possible (there were doubts if seizures started six or more months later). Seizures were generalized in 42% of the cases, and partial or with partial beginning in 45% and multiple types in 13%. One patient began the attacks with status epilepticus. 7% of the patients also had a history of seizures prior to their head trauma. In Table 4 we summarize the time interval between the trauma and seizures.

Discussion

These results help to answer some of the controversies surrounding posttraumatic epilepsy. Generally speaking, from the initial figure of 12% for the incidence of seizures after head trauma, one-third of them had acute seizures, half of these remain with seizures and two-thirds of patients had later seizures. The final incidence of epilepsy is 10%. This figure is similar to those found in previous studies. As it happens, in Jennett's studies[6] no significant difference was found in seizure incidence in children. In our work, however, some controversy may arise because of the follow-up and small number of children studied. Nevertheless in Leviton and Cowan's review[8] the range of the incidence of acute seizues in children is very great even in large series with long term follow-up. These authors conclude that ' Children who had early onset seizures however were more likely to have late seizures than children who did not have early onset seizures'[8]. This statement is also valid in our adult studies.

From our data other conclusions are similar to those previously described. It seems important to bear in mind that acute seizures are more frequently generalized and chronic seizures are more frequently focal. The patients with localized cerebral lesions had an increased risk of epilepsy. This is in agreement with data recently reported by D'Alessandro *et al.*[3]. In their study, all patients with post-traumatic epilepsy had focal brain damage, shown on CT scan[3]. Like other authors, we found a significant number of cases of posttraumatic epilepsy beginning later than 5 or 10 years after the trauma. The reason for this late onset is not yet understood. Recently some studies have emphasized the role of the glial cells in brain scar formation[7]. The glia is strongly involved in the enhancement and spreading of brain potentials. This is still a speculative concept — it is still at the experimental stage — but these findings may be of relevance to our understanding of the origin of these late onset cases of posttraumatic epilepsy. Finally, the severity of trauma, scored on the GCS with prolonged unconsciousness, is not so impressive in predicting posttraumatic epilepsy. In our studies a high percentage of patients with a good score on the GCS (more than 12 points) had posttraumatic epilepsy: 45% of them from our consecutive studies and 50% in the other group. It is interesting to note that in their studies Salazar and coworkers found 'a relatively high proportion of patients (40%) with no initial unconsciousness still had posttraumatic epilepsy'[11]. This is the reason why we can conclude that, as far as it is possible to affirm, the relevance of the initial severity of trauma in our studies is overwhelmed by the relevance of the central (focal or hemispheric) neurological deficits. It is impressive that 65% of patients with seizures, from the first group, and 70% from the second one had central (focal or hemispheric) neurological deficits. This seems to be the most relevant factor in precipitating posttraumatic epilepsy. These findings are retrospectively supported by the studies carried out in veterans from the Vietnam war by Salazar *et al.*[12]. These authors observed posttraumatic epilepsy in 75% of patients with residual hemiparesis and in 86% of patients with aphasia, and their epilepsy was lasting up to 9 and 8 years, respectively.

In conclusion, we can expect a very high probability of posttraumatic epilepsy if central (hemispheric or focal) neurological deficits resulting from diffuse or focal brain lesions, both alone or associated, were present in head trauma patients. This has important clinical and neurosurgical implications for the acute and late management of these patients.

Acknowledgement
We thank Mrs. M. Laura Teixeira for her technical assistance.

References

1. Black P, Shephard RH, Walker AE (1975) Outcome of head trauma: age and post-traumatic seizures. In: Porter R, Fitzsimon DW (eds) Outcome of severe damage to the central nervous system. Ciba Foundation Symposium, Vol 34. Elsevier, Excerpta Medica, North Holland, Amsterdam, pp 215–226
2. Caveness WF, Meirowsky AM, Rish BL, Mohr JP, Kistler JP, Dillon JD, Weiss GH (1979) The nature of posttraumatic epilepsy. J Neurosurg 50: 545–553
3. D'Alessandro R, Ferrara R, Benassi G, Lenzi PL, Sabattini L (1988) Computed tomographic scans in post-traumatic epilepsy. Arch Neurol 45: 42–43
4. Feeney DM, Walker AE (1979) The prediction of posttraumatic epilepsy. A mathematical approach. Arch Neurol 36: 8–12
5. Flint G (1988) Seizures and epilepsy. Br J Neurosurg 2 (3): 419–421
6. Jennett B, Miller JD, Braakman R (1974) Epilepsy after non-missile depressed skull fracture. J Neurosurg 41: 208–216
7. Kimelberg HK, Norenberg MD (1989) Astrocytes. Scientific American 260 (4): 44–52
8. Leviton A, Cowan LD (1981) Methodological issues in the epidemiology of seizure disorders in children. Epidemiol Rev 3: 67–89
9. Majkowski J (1980) Posttraumatic epilepsy: risk factors, familial susceptibility and pharmacologic prophylaxis. In: Canger R, Angeleri F, Penry JK (eds) Advances in epileptology: XII Epilepsy International Symposium. Raven Press, New York, pp 323–329
10. Salazar AM, Jabbari B, Vance SC, Grafman J, Amin D, Dillon JD (1985) Epilepsy after penetrating head injury. I. Clinical correlates: a report of the Vietnam Head Injury Study. Neurology 35: 1406–1414
11. Salazar AM, Grafman J, Vance SC, Weingartner H, Dillon JD, Ludlow C (1986) Consciousness and amnesia after penetrating head injury: neurology and anatomy. Neurology 36: 178–187
12. Weiss G, Caveness WF (1972) Prognostic factors in the persistence of posttraumatic epilepsy. J Neurosurg 37 (2): 164–169
13. Weiss GH, Feeney DM, Caveness WF, Dillon D, Kistler JP, Mohr JP, Rish BL (1983) Prognostic factors for the occurrence of posttraumatic epilepsy. Arch Neurol 40: 7–10

Correspondence: Prof. Dr. A. Martins da Silva, Serviço de Neurofisiologia, Hospital de Santo Antonio, 4000-Porto, Portugal.

Acta Neurochirurgica, Suppl. 50, 52–54 (1990)
© by Springer-Verlag 1990

Epilepsy Following Neurosurgical Intervention

S. A. O'Laoire

National Neurosurgical Centre, Beaumont Hospital, Dublin, Eire and Mater Private Hospital, Dublin, Eire

Summary

The incidence of post-surgical epilepsy has been reported to be very high, and related to the pathological condition, to the surgery itself, and, particularly in the case of aneurysms, to the site of the lesion. Prophylactic anti-convulsant medication has been widely recommended on the basis of the perceived high risk of epilepsy.

A prospective analysis of one hundred consecutive survivors of aneurysm surgery treated in a consistent microsurgical manner was performed to assess the incidence and causation of post-operative seizures. Three patients had a single early post-operative seizure. All three had predisposing features; a previous epileptic history in two, and neurological deficit in the other. Only one patient developed repeated seizures (epilepsy); he had major parenchymal damage.

Seizures did not occur during a two to six year follow up in survivors of aneurysm surgery who did not have a previous epileptic history or a persistent post-operative neurological deficit. The site of aneurysm did not influence the development of epilepsy, and middle cerebral aneurysms were not associated with an increased risk.

The low risk of epilepsy does not justify routine anti-convulsant prophylaxis.

Keywords: Epilepsy; craniotomy; aneurysm; subarachnoid haemorrhage.

Introduction

Since surgeons have ventured to treat intracranial disease, the risk of post-operative epilepsy has been recognised, whether due to the underlying disease, the effects of craniotomy, or both[2].

The literature on craniotomy and the risk of post-operative epilepsy is considerably smaller than the literature on head injury, but the findings are similar. The incidence of epilepsy following craniotomy overall is reported to be between ten and fifty percent[4,7,8,9]. However, the actual incidence varies enormously according to the degree of parenchymal involvement of the disease, the operation, or a combination of the two. Risks as low as 0.5% for sub-frontal approaches to the pituitary gland and third ventricular tumours are reported, to as high as ninety-two percent for intracerebral abcess[4,8].

Even within these categories, there is an enormous variation in the reported incidence of epilepsy, and an important feature in the actual incidence appears to be the presence or absence of parenchymal brain injury[1,4,11]. In that regard, analysis of the incidence of epilepsy following surgical treatment of intracranial arterial aneurysms is of particular value, since patients vary from those with a non-coma producing haemorrhage, undamaged by the haemorrhage or surgery, to those suffering parenchymal injury due to the direct affect of haemorrhage or surgery or both[1,4,11,12,14].

High rates of epilepsy have been reported to be associated with risk factors which include parenchymal brain damage, the presence of an intracerebral haematoma, resection of brain, retraction of brain, persistent major post-operative deficit and to be associated with aneurysms at certain sites such as the middle cerebral artery[1,4,11,14]. Several recent large series have reported a very low incidence of post-operative epilepsy, varying from one to seven percent[3,5,10,12,13,14].

Material and Methods

A prospective study of one hundred consecutive survivors of supratentorial approaches to obliterate an intracranial aneurysm was carried out on patients operated upon by one surgeon between January 1981 and November 1984, to assess the incidence of epilepsy and the effect of anti-convulsant medication. The first sixty-seven patients were treated between January 1981 and January 1983 and received prophylactic anti-convulsants, whereas the subsequent thirty-three patients treated up to November 1984 received none. There was no significant difference in the patient profiles of the two groups, those treated with anti-convulsant and those untreated, in terms of age, coma-producing or non coma-producing haemorrhage, Botterall grade on admission and operation, and location of the aneurysm.

Operation was performed as soon as the patient's condition and operating theatre facilities allowed; sixteen percent were operated upon within three days, thirty-four percent within five days, fifty-six percent within seven days, and eighty-seven percent within two weeks of the presenting haemorrhage. In thirteen percent, operation was performed two to four weeks after the initial bleed, due to delay in patient referral or poor clinical condition on admission.

A total of one hundred and eleven craniotomies were performed to treat one hundred and twenty-two aneurysms, of which one hundred and nineteen were clipped and three wrapped, the latter being small sessile lesions. The pterional approach was used in all cases except for aneurysms of the basilar tip which were approached via the sub-temporal route. Middle cerebral aneurysms were exposed through the sylvian fissure which was opened from medial to lateral. In all cases meticulous microsurgical technique was used with minimal brain exposure, brain retraction, and brain resection. Induced hypotension was not used, and slack operating conditions were obtained by drainage of CSF from the basal subarachnoid cisterns, augmented by ventricular tapping in seventeen cases.

Sub-pial brain resection was avoided whenever possible, and was carried out in twenty-seven patients (twenty-one of thirty-two anterior communicating, five of twenty-two middle cerebral, and one of fifty-seven internal carotid artery aneurysms). In the case of anterior communicating aneurysms, resection was not used to identify the A. 1 segment but confined to exposure of the neck of the aneurysm when necessary.

All survivors were interviewed and examined by the author at 3, 6, 12, 18, and 24 months following surgery and subsequently by annual questionnaire. Specific enquiry was made regarding seizures or any disturbance of consciousness, behaviour or neurological function which might suggest seizures. Follow-up ranged from a minimum of two years to six years.

Results

Ninety-two patients returned to their normal occupation and social life; eighty were without any neurological deficit (excellent result), and twelve had a detectable deficit which did not interfere with function (good result). Five patients were independent but neurological deficits prevented their return to normal activity (fair result), and three patients required assistance with daily living (poor result). There was no difference in the surgical outcomes of the treated and untreated groups.

Three patients had an early seizure following surgery. In the Phenytoin treated group two of sixty-seven developed an early seizure; in the group not receiving anti-convulsant prophylaxis, one of thirty-three had an early seizure. The two patients (excellent result) of the prophylactically treated group both had had seizures earlier in life; one (posterior communicating aneurysm) had been subject to febrile convulsions during childhood illnesses; the other (middle cerebral aneurysm) had seizures after electro-convulsive therapy for depression; a single seizure occurred within twenty-four hours of surgery. In the untreated group one patient (anterior communicating aneurysm), who had a good result (minor neurological deficit), had a seizure two months following surgery. None of those three patients experienced further seizures, and continued on or were placed on anti-convulsant medication for a period of twelve months which was then stopped without recurrence of seizures during the follow-up periods ranging from two to six years. One patient (anterior communicating aneurysm) who had a poor result developed epilepsy two years following surgery. Thus a single early post-operative seizure occured in two patients who were undamaged by the subarachnoid haemorrhage or surgery, but who had a history of earlier seizures. A single early post-operative seizure occurred in one patient who had clinical evidence of parenchymal damage (mild hemiparesis). Repeated seizures (epilepsy) developed two years following surgery in a patient with clinical and CT scan evidence of major parenchymal damage due to angiographically demonstrated arterial spasm.

Discussion

There is now a clear picture emerging that patients who are undamaged by the haemorrhage or surgery have a very low incidence of epilepsy. It appears likely that surgical procedures which minimise brain retraction, and which avoid brain resection in patients who are undamaged by the initial haemorrhage, and who do not subsequently develop the cerebral ischemia associated with progressive vasospasm, have an exceedingly low incidence of epilepsy, which may not be greatly different from the risk of spontaneous epilepsy in the general population[6, 12]. In two studies, post-craniotomy epilepsy, when it occurred, was confined to the post-operative period occurring in three or four percent of patients, and was associated with a previous epileptic history[6, 12].

The risk of epilepsy following surgery for middle cerebral aneurysms has been reported to be as high as thirty percent[1] and thirty-eight percent[4]. The incidence of epilepsy in middle cerebral aneurysms of this series is identical to that of aneurysms at other sites. Other recent reports of the results of aneurysm surgery do not indicate a high risk of epilepsy in the case of middle cerebral aneurysms[3, 5, 12, 13, 14]. A high correlation between middle cerebral aneurysm site, intracerebral haematoma, parenchymal damage, and persistent major post-operative deficits is documented in the reports of high epilepsy risk[1, 4] and it would appear likely that it is those factors rather than the site of the aneurysm which is associated with the high reported rate of epilepsy.

Although the numbers of middle cerebral aneurysms in this series are small, they are sufficient to establish that the incidence of epilepsy following microsurgical approaches to middle cerebral aneurysms through the subarachnoid space is very low, and not higher than aneurysms elsewhere. The parenchymal damage inherent in transcortical approaches to middle cerebral artery aneurysms may be a causative factor in producing high rates of post-surgical epilepsy, and it would appear that the trans-sylvian route is the approach of choice.

The role of prophylactic anti-convulsant medication in situations where the risk of epilepsy is in very low percentage figures must be questioned[12].

References

1. Cabral RJ, King TT, Scott DF (1976) Epilepsy after two different neurosurgical approaches to the treatment of ruptured intracranial aneurysm. J Neurol Neurosurg Psychiatry 39: 1052–1056
2. Cushing H, Eisenhardt L (1915) Meningiomas. C Thomas, Springfield Ill, p 875
3. Fabinyi GCA, Artiola-Fortuny L (1980) Epilepsy after craniotomy for intracranial aneurysm. Lancet June 14, 1299–1300
4. Foy PM, Copeland GP, Shaw MDM (1981) The incidence of post-traumatic seizures. Acta Neurochir (Wien) 55: 253–264
5. Krayenbuhl HA, Yasargil MG, Flamm ES et al (1972) Micro-surgical treatment of intracranial saccular aneurysms. J Neurosurg 37: 678–686
6. Kvam DA, Loftus CM, Copeland BR, Quest DO (1982) Seizures during the immediate postoperative period. Neurosurgery 12: 14–17
7. Matthew E, Sherwin AL, Welner SA et al (1980) Seizures following intracranial surgery: Incidence in the first post-operative Can J Neurol Sci 7: 285–290
8. North JB, Penhall RK, Hanieh A (1980) Post-operative epilepsy: A double-blind trial of phenytoin after craniotomy. Lancet Feb 23, 384–386
9. North JB, Penhall RK, Hanieh A (1983) Phenytoin and post-operative epilepsy: A double-blind study. J Neurosurg 58: 672–677
10. Ropper Ah, Zervas NT (1984) Outcome 1 year after SAH from cerebral aneurysm. J Neurosurg 60: 909–915
11. Rose FC, Sarner M (1965) Epilepsy after ruptured intracerebral aneurysm. Br Med J 1: 18–21
12. Sbeih I, Tamas LB, O'Laoire SA (1986) Epilepsy after operation for aneurysms. Neurosurgery 19: 784–787
13. Shephard RH (1983) Ruptured cerebral aneurysms: Early and late prognosis with surgical treatment. J Neurosurg 59: 6–15
14. Sundt TM, Kobayashi S, Fode N et al (1982) Results and complications of surgical management of 809 intracranial aneurysms in 722 cases. J Neurosurg 56: 753–765

Correspondence: S. A. O'Laoire, National Neurosurgical Centre, Mater Private Hospital, Eccles Street, Dublin 7, Eire.

Acta Neurochirurgica, Suppl. 50, 55–57 (1990)
© by Springer-Verlag 1990

Post-operative Epilepsy and the Efficacy of Anticonvulsant Therapy

M. D. M. Shaw

Mersey Regional Department of Medical and Surgical Neurology, Walton Hospital, Liverpool, U.K.

Summary

Several supratentorial pathologies, including anterior/middle cerebral artery aneurysms, arteriovenous malformations, spontaneous intracerebral haematomas, meningiomas and abscesses, are associated with a high incidence of post-operative epilepsy. Prophylactic anticonvulsant therapy does not significantly reduce this incidence.

Keywords: Post-operative seizures; post-operative epilepsy; prophylactic anticonvulsants.

Introduction

The risk of developing epilepsy following supratentorial procedures for non traumatic pathology has been recognised for many years. In the booklet on medical aspects of fitness to drive, published by the Medical Commission on Accident Prevention, Bond and his coauthors named neurosurgery as one of the conditions in which they considered that epilepsy was such a frequent complication that it constituted a prospective disability with respect to a person being licensed to drive in the United Kingdom[1]. As we did not know the actual incidence of post-operative epilepsy in the various supratentorial procedures, it was decided to undertake a retrospective study in order to ascertain the true incidence occurring over a relatively long period of time[2], together with any particular factors which might increase the risk of developing post-operative seizures. Ideally a prospective study would have been preferable but at least in the Mersey Regional Unit this had become impossible because the surgeons were increasingly using prophylactic anticonvulsants.

Material and Methods of the Retrospective Study on Post-operative Epilepsy Incidence

During the 5 year period under study, namely 1970–1974 inclusive, 1102 patients were admitted to the Regional Neurosurgical Unit at Walton Hospital and underwent consecutive supratentorial neurosurgical procedures for non-traumatic conditions. The minimum period of follow up was 5 years unless death intervened earlier. A thousand patients were successfully traced[2]. The incidence of seizures was determined by review of the case histories, outpatient follow up and by contacting the general practitioner. It is very likely, because of the patient's reluctance to admit to seizures, that the incidences defined are an underestimate[2,3].

Results of the Study on Post-operative Epilepsy Incidence

91 patients underwent a direct surgical approach for vascular lesions. There were 252 aneurysms, 24 arteriovenous malformations and 15 spontaneous intracerebral haematomas (Table 1). In the case of aneurysms the presence of a haematoma increased the risk to 42%. None of the 291 patients undergoing surgery for vascular conditions had a prior history of epilepsy.

61 patients underwent surgery for meningioma of which 11 were situated on the convexity, 15 parasagittally and 35 at the base. The overall incidences of epilepsy are shown in Table 2. In those patients who had had pre-operative epilepsy 44% had no seizures post-operatively. There was no significant difference in this regard between the various sites. The risk of developing epilepsy de novo was 22%.

338 patients underwent surgery for a supratentorial malignant tumour; 301 were gliomas of various degrees

Table 1. *Vascular Lesions*

	Seizure incidence (%)
Overall	22
AVM	50
Spont. haematoma	20
Aneurysm:	
Anterior complex	21
Middle cerebral	38

Table 2. *Neoplasms*

	Seizure incidence (%)	
	pre-op.	post-op.
Meningioma	41	36
Glioma	28	
Biopsy		9
Resection		20
Metastasis	28	
Biopsy		9
Removal		20

Table 3. *Other Pathologies*

	n	Seizure incidence (%)
Suprasellar lesions	40	5
csf internal shunt	57	22
Abscess	13	92
Ventriculography	100	14
Frazier's op. *	63	3

* Vth. nerve root section Meckel's cave.

of differentiation, and 37 were metastases. The difference in the incidence post-operatively (Table 2) between biopsy and craniotomy is not significant.

The incidences of post-operative seizures in other supratentorial pathologies are shown in Table 3. In those undergoing ventriculography, we were unable to find any correlation between the incidence of epilepsy and the contrast medium used or the pathology of the condition leading to the investigation[2].

It was important to define the time scale of the risk of post-operative epilepsy and whether this was related to the type of seizure[3]. Only the 877 patients who had had no prior history of epilepsy were considered. 57% of these patients survived for more than 5 years. The type of seizure was classified as either focal, including motor sensory and temporal lobe epilepsy, grand mal epilepsy or a combination of both. A single seizure was defined as either grand mal or focal seizure occurring in isolation. If focal seizure became generalised it was accepted as being a single seizure. Jennett's definition of early seizures as those occurring within the first 7 days and his late definition as those occurring after one week[4] were adopted.

Early seizures were found in only 6% of all patients who underwent supratentorial surgery but of the pa-

tients who eventually developed post-operative seizures 37% experienced a seizure within 7 days of surgery. At one and two years post-operatively 77% and 92% respectively of all patients who developed seizures had experienced their first. There was no significant difference between the time of onset of the seizures (*i.e.* early/late) and the type of surgery or the condition for which it was undertaken, with the exception of ventriculography in which 71% of cases had the first seizure within 7 days of the investigation. The only patients who had a continuing risk beyond 5 years were those with an abscess. In the vascular group no patient who underwent surgery for aneurysm developed post-operative seizures later than 2 years after surgery. However in the arteriovenous malformation and the spontaneous intracerebral haematoma patients the risk of onset at 2 years was still 14%.

Of the 37 patients having early post-operative seizures 41% went on to experience late seizures. All 5 patients who underwent surgery for supratentorial abscess and who experienced early seizures developed late epilepsy but of those who developed early fits after ventriculography only 20% continued to experience seizures. In the vascular group an early onset of seizures was more commonly associated with surgery for anterior cerebral artery aneurysm that it was for aneurysms in other positions though the difference did not reach statistical significance. However, only 13% of the patients who underwent surgery for an aneurysm at this site and who developed an early seizure subsequently developed late seizures, whereas all the patients who underwent surgery for an arteriovenous malformation or for a spontaneous intracerebral haematoma continued to experience late seizures.

The type of seizures showed no correlation with either the time of onset or the type of surgery performed. However, 70% of patients in whom the early seizures had a focal element preceding a grand mal seizure continued to experience epilepsy. Single seizures occurred in 21% of the 146 patients who had post-operative seizures. A higher incidence of single seizures was found in the early post-operative period, 48%, compared to 5% occurring in the period after one week. The incidence of single seizures varied from surgical group to surgical group. Single seizures were common after ventriculography but did not occur after surgery for abscess or meningioma.

The highest incidence of post-operative epilepsy occurs within the first 3 months but it is not until 6 months have elapsed that the overall risk falls below 10%. However in those patients who have undergone surgery for middle cerebral artery aneurysms this level of risk is not reached until 2 years and between 2 and 5 years in the arteriovenous malformation and spontaneous intracerebral haematoma groups[3].

Thus *the conclusions from these studies*[2, 3] were that the overall risk of post-operative epilepsy was 17%, that certain groups carry a much higher risk and that, though the overall incidence dropped to 10% by six months, in some conditions this level of risk continued for some years.

Prospective Study on Efficacy of Anticonvulsant Therapy

Material

A prospective study was therefore undertaken to determine whether prophylactic anticonvulsants would control post-operative seizures. High risk (i.e. at least 20%) patients were admitted to the study (Table 4). Because other authors[5] had recorded much higher incidences of post-operative epilepsy following subfrontal exploration for pituitary tumours, these patients were included together with other benign pathologies arising in the midline such as craniopharyngiomas. Patients with a history of epilepsy were excluded. Patients were randomised to one of 3 treatments, carbamazepine, phenytoin or no treatment. Each of the active arms was divided into 6 months or 24 months therapy. The occurrence of seizures was treated as an end point to this part of the study and any anticonvulsant therapy could then be instituted by the clinician. These patients were however followed up to learn whether seizures remained a continuing problem. Follow up was 2–7 years in duration. Amongst the survivors few were lost to follow up.

Results

The overall seizure rate in the prospective study was 36% compared with 34% for the high risk groups, in the retrospective studies[2, 3].

The incidence of seizures, in the prospective study,

Table 4. *High Risk Conditions*

Aneurysm
    anterior cerebral
    middle cerebral

Arteriovenous malformation
Spontaneous intracerebral haematoma
Meningioma
Abscess
Suprasellar lesions

whilst on/off therapy was 42% and 34% for those on carbamazepine for 6 and 24 months respectively, 38% and 34% for phenytoin for 6 and 24 months, and 39% in the no treatment group. The apparent effect of carbamazepine in the 6 month group did not reach significance, nor was it substantiated when those patients on carbamazepine for 24 months were looked at 6 months and added to this group. Furthermore in this study, unlike traumatic epilepsy, the presence of early seizures does not seem to influence the incidence of late seizures. North *et al.*[6] reported that phenytoin used prophylactically altered the incidence of seizure between day 7 and day 70. We therefore determined the seizures incidence occurring between 1–13 weeks and found that neither anticonvulsant influenced this incidence occurring within this period.

There was a suggestion that those patients who were compliant did have a lower incidence of post-operative seizures when compared with the control group but the difference did not reach statistical significance. Side effects, and in particular skin rashes did occur in a high percentage of patients.

Considering those patients who survived for at least 2 years, 36% had had more then one seizure during the last year of their follow up. Considering the last 2 years of follow up 55% were found to have had more than one seizure. Post-operative epilepsy is therefore a continuing problem for many patients.

Thus *in conclusion* prophylactic anticonvulsants in post-operative epilepsy seem to be ineffective and are associated with a considerable incidence of side effects.

References

1. Raffle A (ed) (1985) Medical aspects of fitness to drive. A guide for medical practitioners. Fourth edition. The Medical Commission on Accident Prevention
2. Foy PM, Copeland GP, Shaw MDM (1981) The incidence of post-operative seizures. Acta Neurochir (Wien) 55: 253–264
3. Foy PM, Copeland GP, Shaw MDM (1981) The natural history of post-operative seizures. Acta Neurochir (Wien) 57: 15–22
4. Jennett B (1975) Epilepsy after non-missile head injuries. Second edition. Heinemann
5. Cast IP, Wilson PJ (1981) Pituitary tumours. A ten year survey. J Neurol Neurosurg Psychiatry 44: 371
6. North JB, Penhall RK, Hanieh A, Frewin DB, Taylor WB (1983) Phenytoin and post-operative epilepsy. J Neurosurg 58: 672–677

Correspondence: M. D. M. Shaw, Mersey Regional Department of Medical and Surgical Neurology, Walton Hospital, Rice Lane, Liverpool, L 9 1AE, U.K.

Acta Neurochirurgica, Suppl. 50, 58–63 (1990)

Principles of Surgery for Epilepsy*

G. F. Rossi

Institute of Neurosurgery, Catholic University, Rome, Italy

Summary

The pre-requisite conditions for surgical treatment of epilepsy are: ineffective pharmacotherapy, the unacceptable nature of the seizures and the presence of an organic brain lesion as the basis of the disease. The principles of surgery stem from knowledge of the anatomico-functional structure and evolution of the epileptic process, which indicate the targets for surgery. Two main groups of surgical approaches are available. The first and most efficient one aims at suppressing the seizures by acting on the nucleus of origin of the epilepsy; the combined removal of both the causative cerebral lesion and the primary epileptogenic zone is considered as the "optimal" type of surgery (hemispherectomy, lobectomy, topectomy); only in rare cases is an ablation limited either to the causative lesion (mainly in children) or to the epileptogenic zone sufficient. The second group of surgical procedures aims at reducing the cerebral epileptogenicity by preventing diffusion of the epileptic discharges (callosotomy, subpial transections), by enhancing inhibitory (cerebellar stimulation) or reducing facilitatory influences (stereotactic deep lesions); these are regarded as a "second choice" treatment. The final surgical indication, the choice of the surgical approach and the surgical prognosis are dependent on accurate presurgical investigations. The very good results which can be obtained should favour the wider use of surgery.

Keywords: Epilepsy; surgical treatment; principles; indications; results.

I. Introduction

Epilepsy is a multifactorial disease. It results from the concurrence of pathogenetic factors of a different nature. Most of them are common to all patients. However, each factor can vary in importance in the different patients. Likewise, the resulting clinical syndrome, though showing some common basic characteristics, can be quite multifarious, both with regard to its qualitative aspects and its severity. The general principles of surgery for epilepsy are grounded upon the knowledge of the basic aspects of the disease; their practical application requires a detailed knowledge of the characteristics of the epileptic process in each patient.

II. The Logical Basis of Surgery

There are certain *preliminary conditions* to surgery for epilepsy:

i) the resistance of the epileptic syndrome to the pharmacological treatment,

ii) the unacceptable character (quality, severity, frequency) of the seizures, and

iii) the origin of epilepsy from an organic brain lesion (therefore excluding the genetically determined epilepsies and those due to extracerebral factors, such as for instance, metabolic disorders).

Any rationale for surgery starts from these conditions. Then, the construction of the logical basis of surgery for epilepsy proceeds according to our knowledge of the *anatomico-functional structure and evolution of the epileptic process*. Even if, as mentioned above, several different pathogenetic factors are at play, a common basic line of development can be suggested.

The first step is the occurrence of a lesion of the cerebrum. This is the event which actually starts the process of epileptization. The nature of the lesion is not specific; its size is not important; its location however, can be relevant. Let us assume that we are dealing with a stabilized, not evolving lesion.

The second step is the occurrence of local structural, circulatory and metabolic changes as a consequence of the above mentioned lesion. Some neurons disappear, others show regressive changes; the glial cells proliferate[1, 23, 28, 40]. The capillary network is distorted

---

* The reported personal findings were obtained in researches partially supported by the Ministero della Pubblica Istruzione and the Consiglio Nazionale delle Ricerche.

and reduced[1, 23]; local blood flow and local metabolism decrease[2, 13]. The local neurochemical environment is altered; of particular interest is the decrease in inhibitory neurotransmitters[18]. All these events are obviously interdependent.

The third step is the functional impairment of the affected cerebral region as a consequence of the above. A reduction in function can be found. It is a nonspecific event. Its clinical connotation and relevance are dependent on the extension and functional properties of the affected cerebral region. What interests us most here is the occurrence of epileptization (primary epileptization). This second type of neuronal functional abnormality does not become manifest in all patients, but only in some of them. It is probably related to the combination of peculiar phenomena developing in the vicinity of the causative lesion (mentioned in the second step) with an inborn, genetically determined factor, the so-called "predisposition" to epilepsy[16]. Typical characteristics of the activity of the primarily epileptized cerebral neurons are the tendency to discharge in bursts and in synchrony[40, 43], as well as the scarce sensitivity to internal and external influences with a consequently marked stability[29, 32, 40, 43, 44].

The fourth step is the tendency of the neuronal epileptization to expand, if not controlled. The phenomenon is due to the recruitment of new neurons, exposed to the continuous impinging upon them of the abnormal impulses arising from the primarily epileptized ones (secondary epileptization). The secondarily epileptized neurons are usually located in the vicinity of the primarily epileptized ones but can be found also at a distance from them, in the same hemisphere and in the contralateral one. Their epileptization is dependent on that of the primary epileptized neurons. Unlike primary epileptization, the secondary one is quite variable, because it is easily affected by influences of a different nature (facilitating or inhibiting influences[29, 32, 40, 44]). When the total mass of epileptized neurons (i.e. primary plus secondary) reaches a certain, critical level, their synchronous discharge can give origin to a clinically manifested seizure.

Finally, the fifth step. Some events can lead to an evolution of the epileptic disease. The most important of them is the occurrence or the extension of neuronal damage because of repeated seizures[4, 5, 10]. The autonomization of the secondary epileptization (the "mirror focus"[20]), as well as the enhancement of epileptogenicity in cerebral regions under the influence of the primary epileptized cerebral zone through a "kindling"-like mechanism, have been suggested[12, 20], though not

by all[14]. Impairment of cerebral functional capacities and behavior can thus appear or worsen[3, 10]. The often necessary, heavy drug therapy can contribute towards this[6, 9, 39].

Obviously, these steps in the development and evolution of the process of epileptization do not occur in all epileptic patients. On the contrary, in the majority of cases the pharmacological treatment is apt, if not to suppress, at least to stop the process when in its initial stages. The full development of such a process through all or most of its five steps characterizes those patients who are considered for surgery.

III. The Targets for Surgery and the Surgical Approaches

This proposed schematization of the most important aspects of the anatomico-functional structure and evolution of the epileptogenic process permits the identification of possible targets for surgery; i) the causative cerebral pathology, ii) the primary epileptized cerebral zone, iii) the pathways of propagation of the epileptic discharges from the primary epileptogenic zone to other cerebral districts, iv) the influences facilitating or inhibiting the cerebral epileptogenicity, and v) the circulatory, metabolic and neurochemical abnormalities of the cerebral region close to the causative lesion.

Given these targets, the following basic *surgical approaches and techniques* can be considered in order to treat the epileptic disease.

1. The first approach is the removal of the causative cerebral pathology. Actually such a surgical modality should not be included among the techniques of "surgery for epilepsy", on account of the fact that it does not act directly on the epileptogenic cerebral neurons, which are left intact. However, there is no doubt that in certain cases it is followed by the disappearance of epilepsy[10, 34]. In our experience[34], this occurs mainly in children and when epilepsy has been of short duration. The main limitation of this surgical procedure is that it does not permit us to foresee whether and to what extent epilepsy will be affected. On the other hand, it has the advantage of sparing the cerebral tissue and, if the epileptological result is not good, of permitting the removal of the epileptogenic cerebral zone with a second operation later on. For these reasons, we regard this surgical modality as a possible "peliminary" approach[30].

2. The second surgical approach aims at removing the ensemble of the cerebral lesion, plus the local structural abnormalities, plus the primary epileptized cer-

ebral zone (namely, what we have defined as the lesional-functional epileptogenic complex[35]). It has to be regarded as the most classical of the surgical approaches to epilepsy. Obviously, it is necessary that both the causative lesion and the primary epileptogenic zone be clearly located and safely accessible to surgery. The most utilized surgical techniques are topectomy, lobectomy and hemispherectomy; the selective removal of deeply located lesional-functional epileptogenic complexes (such as the selective amygdalohippocampectomy, to be discussed in a subsequent report) can be included among these.

3. The third approach can be considered when the cerebral zone containing the location of the primarily epileptized neurons (i.e. the epileptogenic zone) has been defined and is compatible with surgical removal, but there is no evidence of the causative cerebral lesion (as might happen in certain cases of gliosis). In these cases, surgery is limited to the removal of the epileptogenic zone.

4. The fourth approach aims at preventing or limiting the propagation of the epileptic discharges from their site of origin to other cerebral districts or, in cases of multifocal epilepsy, from one epileptogenic zone to another. The best known surgical technique is partial or complete section of the largest of the cerebral commissures, namely the corpus callosum[26,38]. Another technique, even though very rarely performed, utilizes multiple subpial transection[21]. These surgical approaches can be considered when there is no clearcut evidence of a well localized and surgically accessible lesional-functional epileptogenic complex, and when secondary generalization is an important aspect ot the epileptic syndrome, as well as in certain multifocal epilepsies.

5. The fifth surgical approach attempts to reduce the level of cerebral epileptogenicity either by enhancing inhibitory or by reducing facilitating influences. Electrical stimulation of the anterior lobe of the cerebellum is the best known of the techniques used for the first purpose[8,33]. Caudate stimulation has been also proposed[22]. Stereotactic lesioning of several subcortical sites has been performed to fulfill the second purpose (the fields of Forel and the amygdala being the preferred targets[25,36]). Both the fourth and fifth surgical approaches might be taken into consideration when the first three cannot be used.

6. Finally, it is worth mentioning another modality of surgical treatment, that is still at an experimental stage. Its purpose is to reduce the neuronal epileptization by correcting the local neurochemical environment. Local chronic infusion of neurotransmitters, or the local implant of live tissue producing them might be utilized. GABA-agonist or GABA-mimetic substances and certain catecholamines appear to be particularly interesting[19,22,27,37,41].

IV. The Choice of the Surgical Approach: The Presurgical Diagnostic Investigations

The possibility of surgically treating an epileptic patient as well as choosing the most convenient surgical approach are dependent on knowledge of the many aspects of the epileptic disease in each individual patient. Such knowledge can be provided by an accurate presurgical diagnostic work-up.

Several different clinical, neuroradiological, isotopic and electrophysiological procedures are available today[11,42]. Some of them are apt to provide information about the causative lesion (CT scan, NMR), others about the related nearby structural, circulatory and metabolic abnormalities (rCBF, PET, SPECT), and others about the consequent neuronal function disorder (neurological and neuropsychological examinations, analysis of the clinical seizure manifestations, study of the background and epileptic electrical cerebral activity). Some of these procedures require a high specialization, as for instance those based on the short or long-term recording of electrical cerebral activity directly from the brain, through electrodes stereotactically implanted in preselected cerebral sites or applied to the brain surface. Others are very expensive and not easily available, as for instance PET. Not all the diagnostic investigations mentioned above have to be used in all patients. The choice is made on the basis of the characteristics of the epileptic disease of the single patient. We want to stress, however, that all or at least most of them should be directly or indirectly available in all centres involved in surgery for epilepsy.

V. The Results of Surgery

When commenting on the results of surgery one usually refers to the effect of the surgical treatment on the seizures, which is the most obvious manifestation of the epileptic disease. The seizures can disappear, or can be reduced in number and severity, or can remain as they were before surgery; deterioration is possible, but very rare. It is commonly agreed that a reliable evaluation can be made only after at least two years from the operation, while continuing on pharmacological treatment. Attempts to reduce and then to stop the drug treatment in cases of total seizure disappearance can be initiated after this period.

Table 1. *Removal of the Lesional-Functional Epileptogenic Complex* (minimum follow-up: 2 years)

	Cases	Complete seizure disappearance
Hemispherectomy	11	10 (91%)
Lobectomy	63	38 (60%)
Topectomy	33	15 (46%)
Total	107	63 (59%)

There is no doubt that the best chances of obtaining the optimal results are provided by the surgical approach aimed at removing both the causative lesion and the primary epileptized cerebral zone (see III, 2 above). If the classification of the results is simplified by showing on one side complete abolition of seizures, and on the other side the persistence of seizures (even if reduced in number and/or severity), this type of surgery appears to give the optimal result in 50 to more than 80 percent of cases, depending on several factors[7,11,15,24,31]. Our personal experience is in line with these findings. A relation between the result and the surgical technique utilized (topectomy, lobectomy, hemispherectomy) is apparent: the larger the surgical removal, the better the result (see Table 1).

This finding is compatible with other data indicating the importance, as well as the difficulties, of spatially delineating the epileptogenic cerebral zone[31]. Good results can also be provided, though definitely to a lesser extent, by other surgical modalities, and in particular by the removal of the epileptogenic zone only (see III, 3) and by the longitudinal section of the corpus callosum (see III, 4). The latter technique is certainly interesting, but the problem of the selection of candidates is still debated[26,38]. The evidence available so far does not appear sufficient for a safe evaluation of the results of multiple subpial transection[21] (see III, 4), cerebellar electrical stimulation[33] and subcortical stereotactic lesioning[25,36] (see III, 5).

In our opinion, a clear distinction should be made between the surgical approach aiming at removing both the causative lesion and the epileptogenic zone, and all the others: the former is the "optimal" surgical approach, and the others belong to a "second choice" type of surgery, to be considered when the former cannot be applied[30].

VI. Concluding Remarks

Surgery for epilepsy is, or should be, based on principles which stem from a knowledge of the clinical syndrome, and of the anatomico-functional structure of the underlying epileptogenic process. Its results are dependent on several factors. Although they obviously share several basic elements, epilepsies can show peculiar aspects in different patients. These aspects can be very important, firstly in planning the presurgical investigations, then in deciding the final indication for surgery and the choice of the appropriate surgical approach, and finally in formulating the surgical prognosis.

We believe that the strategy of the presurgical and surgical work has to be to a great extent "personalized". As we have stressed elsewhere[30], standardization of the surgical treatment of epilepsy is hardly compatible with the principles dicussed here.

This presurgical and surgical work requires a complex organization, necessitating money, expert and motivated people, the availability of many diagnostic investigations and the possibility of performing the different techniques of surgery.

There is no doubt that the results that surgery can offer to epileptics, not helped sufficiently by pharmacological treatment, can be very good. Nevertheless, surgery is scarcely utilized[17,22,30,41]. This underutilization, to be discussed in another report, might at least in part depend on the difficulty in establishing the complex organization mentioned above. In our opinion, however, it is likely to be related as much to the scarce (and sometimes distorted) information reporting the surgical possibilities available today.

References *

1. Babb TL, Brown WJ (1986) Neuronal, dendritic and vascular profiles of human temporal lobe epilepsy correlated with cellular physiology in vivo. In: Delgado-Escueta AV *et al* (eds) Advances in neurology, Vol 44. Raven Press, New York, pp 949–966
2. Baldy-Moulinier M, Ingvar DH, Meldrum BS (eds) (1983) Current problems in epilepsy 1. John Libbey, London Paris, 400 pp
3. Bjørnaes H (1988) Consequences of severe epilepsy: psychosocial aspects. In: Dam M *et al* (eds) Surgical treatment of epilepsy. Acta Neurol Scand [Suppl 117] 78: 28–32
4. Bruton CJ (1987) Pathological findings caused by seizures, with particular reference to the temporal lobe. In: Wieser HG, Elger CE (eds) Presurgical evaluation of epileptics. Springer, Berlin Heidelberg New York, pp 71–78
5. Dam AM (1988) Consequence of severe epilepsy: Neuropathological aspects. In: Dam M *et al* (eds) Surgical treatment of epilepsy. Acta Neurol Sccand [Suppl 117] 78: 24–27
6. Dam M (1988) Side-effects of drug treatment in epilepsy. In: Dam M *et al* (eds) Surgical treatment of epilepsy. Acta Neurol Scand [Suppl 117] 78: 34–40

* On account of the general character of the present report, most of the references are related to review articles or books.

7. Dam M, Gram L, Schmidt K (eds) (1988) Surgical treatment of epilepsy. Acta Neurol Scand [Suppl 117] 78

8. Davis R, Bloedel JR (eds) (1984) Cerebellar stimulation for spasticity and seizures. CRC Press, Boca Raton

9. Dreifuss FE (1983) Adverse effects of antiepileptic drugs. In: Ward AA jr *et al* (eds) Epilepsy. Raven Press, New York, pp 249–266

10. Dreifuss FE (1987) Goals of surgery for epilepsy. In: Engel J jr(ed) Surgical treatment of the epilepsies. Raven Press, New York, pp 31–49

11. Engel J jr (ed) (1987) Surgical treatment of the epilepsies. Raven Press, New York

12. Engel J jr, Cahan LD (1986) Potential relevance of kindling in human partial epilepsy. In: Wada JA (ed) Kindling 3: Raven Press, New York, pp 37–51

13. Engel J jr, Cahan LD, Sutherling WW, Crandall PH, Phelps ME (1987) The use of positron emission tomography in the surgical treatment of epilepsy. In: Wieser HG, Elger CE (eds) Presurgical evaluation of epileptics. Springer, Berlin Heidelberg New York, pp 136–139

14. Goldensohn ES (1984) The relevance of secondary epileptogenesis to the treatment of epilepsy: kindling and the mirror focus. Epilepsia 25 [Suppl 2]: 5156–5168

15. Green JR, Sidell AD (1982) Neurosurgical aspects of epilepsy in children and adolescents. In: Youmans JR (ed) Neurological surgery, vol 6. Saunders, Philadelphia, pp 3858–3909

16. Houser Wa, Annegers JF, Anderson VE (1983) Epidemiology and the genetics of epilepsy. In: Ward AA jr *et al* (eds) Epilepsy. Raven Press, New York, pp 267–294

17. Janz D (1987) Consequence for the present practice of epilepsy therapy in Europe. In: Wieser HG, Elger CE (eds) Presurgical evaluation of epileptics. Springer, Berlin Heidelberg New York, pp 373–375

18. Lloys KG, Bossi L, Morselli PL, Munari C, Rougier M, Loiseau H (1986) Alterations of GABA-mediated synaptic transmission in human epilepsy. In: Delgado-Escueta AV *et al* (eds) Advances in neurology, vol 44. Raven Press, New York, pp 1033–1044

19. Meldrum BS (1986) Pharmacological approaches to the treatment of epilepsy. In: Meldrum BS, Porter RJ (eds) Current problems in epilepsy 4. John Libbey, London Paris, pp 17–30

20. Morrell F, Wada J, Engel J jr (1987) Appendix III: potential relevance of kindling and secondary epileptogenesis to the consideration of surgical treatment for epilepsy. In: Engel J jr (ed) Surgical treatment of the epilepsies. Raven Press, New York, pp 701–707

21. Morrell F, Whisler WW, Bleck TP (1989) Multiple subpial transection: a new approach to the surgical treatment of focal epilepsy. J Neurosurg 70: pp 231–239

22. Ojemann GA (1987) Surgical therapy for medical intractable epilepsy. J Neurosurg 66: pp 489–499

23. Paul LA, Scheibel AB (1986) Structural substrates of epilepsy. In: Delgado-Escueta AV *et al* (eds) Advances in neurology, vol 44. Raven Press, New York, pp 775–786

24. Purpura DP, Penry JK, Walter RD (eds) (1975) Advances in neurology, vol 8. Raven Press, New York, 356 pp

25. Ravagnati L (1987) Stereotactic surgery for epilepsy. In: Wieser HG, Elger CE (eds) Presurgical evaluation of epileptics. Springer, Berlin Heidelberg New York, pp 361–371

26. Reeves AG (ed) (1985) Epilepsy and the corpus callosum. Plenum Press, New York London, 534 pp

27. Ribak CE (1983) Morphological, biochemical, and immunocytochemical changes of the cortical, GABAergic system in epileptic foci. In: Ward AA jr *et al* (eds) Epilepsy. Raven Press, New York, pp 109–130

28. Robitaille Y (1987) Neuropathological findings in epileptic foci. In: Wieser HG *et al* (eds) Current problems in epilepsy 3. John Libbey, London Paris, pp 95–112

29. Rossi GF (1973) Problems of analysis and interpretation of electrocerebral signals in epilepsy. A neurosurgeon's view. In: Brazier MAB (ed) Epilepsy. Its phenomena in man. Academic Press, New York, pp 259–285

30. Rossi GF (1988) Trattamento chirurgico delle epilessie farmacoresistenti: inquadramento generale. Boll Lega It Epil 64: 35–39

31. Rossi GF, Colicchio G, Gentilomo A, Scerrati M (1978) Discussion on the causes of failure of surgical treatment of partial epilepsies. Appl Neurophysiol 41: 29–37

32. Rossi GF, Colicchio G, Pola P (1984) Interictal epileptic activity during sleep: a stereo-EEG study in patients with partial epilepsy. EEG Clin Neurophysiol 58: 97–106

33. Rossi GF, Colicchio G, Scerrati M (1987) La stimolazione cerebellare cronica nel trattamento dell' epilessia farmacoresistente. Boll Lega It Epil 60: 5–12

34. Rossi GF, Di Rocco C, Iannelli A (1988) The role of the neurosurgeon in the treatment of childhood epilepsy. In: Faienza C, Prati GL (eds) Excerpta Medica Intern. Cong. Ser. 781. Elsevier, Amsterdam New York, pp 135–147

35. Rossi GF, Gentilomo A, Colicchio G (1974) Le problème de la recherche de la topographie d'origine de l'épilepsie. Schweiz Arch Neurol Neurochir Psychiatr 115: 229–270

36. Scerrati M (1989) La chirurgia stereotassica nel trattamento delle epilessie farmacoresistenti. Boll Lega It Epil 64: 97–101

37. Spencer DD (1987) Postscript: should there be a surgical treatment of choice, and if so, how should it be determined? In: Engel J jr (ed) Surgical treatment of the epilepsies. Raven Press, New York, pp 477–484

38. Spencer SS (1988) Corpus callosum section and other disconnection procedures for medically intractable epilepsy. Epilepsia 29 [Suppl 2]: S 85–S 99

39. Trimble MR, Thompson PJ (1983) Anticonvulsant drugs, cognitive function and behavior. Epilepsia 24 [Suppl 1]: 555–563

40. Ward AA jr (1975) Theoretical basis for surgical therapy of epilepsy. In: Purpura JK *et al* (eds) Advances in neurology, vol 8. Raven Press, New York, pp 23–35

41. Ward AA jr (1983) Perspectives for surgical therapy of epilepsy. In: Ward AA jr *et al* (eds) Epilepsy. Raven Press, New York, pp 371–400

42. Wieser HG, Elger CE (eds) (1987) Presurgical evaluation of epileptics. Springer, Berlin Heidelberg New York

43. Wieser HG, Speckmann EJ, Engel J jr (eds) (1987) Current problems in epilepsy 3. John Libbey, London Paris

44. Wyler AR, Ward AA jr (1986) Neuronal firing patterns from epileptogenic foci of monkey and human. In: Delgado-Escueta AV *et al* (eds) Advances in neurology, vol 44. Raven Press, New York, pp 967–989

Correspondence: Prof. Dr. G. F. Rossi, Institute of Neurosurgery, Catholic University, Largo Agostino Gemelli, I-00168 Rome, Italy.

Editor's Note

The success of surgery for intractable epilepsy is judged generally against the background of a substantial period of proven failure of drug therapy, appropriately monitored. However, it is essential and ethically feasible with some procedures to design double-blind placebo controlled studies of surgical intervention. The efficacy of cerebellar stimulation for severe intractable epilepsy has provoked considerable controversy and a model double-blind trial was successfully mounted in Southampton. Twelve patients with severe intractable epilepsy were treated by chronic cerebellar stimulation under double-blind conditions for six months. No reduction in seizure frequency occurred that could be attributed to stimulation, though 11 of the patients considered that the trial had helped them. One patient experienced fewer episodes of incontinence during stimulation. Cerebellar stimulation in its present form cannot be recommended for the treatment of severe intractable epilepsy.

Wright GDS, McLellan DL, Brice JG (1984) J Neurol Neurosurg Psychiatry 47: 769–774

Acta Neurochirurgica, Suppl. 50, 64–71 (1990)

Selection Criteria for Surgery – Clinical Features and EEG Including Invasive Techniques

H. G. Wieser

Department of Neurology, University Hospital Zürich, Switzerland

Summary

Preconditions, selection criteria for surgical treatment of epilepsy and some aspects of the presurgical evaluation are discussed. Identification of those patients who will profit from epilepsy surgery, exact localization of the seizure generating structures, and evaluation of the potential risks of the proposed operation are the main goals of the preoperative diagnostic procedures. The localizational value of certain seizure patterns and ictal clinical signs and symptoms is reviewed and illustrated with symptoms provoked by electrical stimulation of the brain.

Keywords: Epilepsy surgery; presurgical evaluation; selection criteria; localizational value of ictal clinical features.

Selection Criteria

The decision whether surgical treatment should be offered a patient suffering from epileptic seizures depends on several factors:

(1) Etiology. Can an underlying cause be detected which demands surgical treatment in its own right? One example is a rapidly growing brain tumour. The opposite situation would be seizures due to a diffuse metabolic, inflammatory, or vascular disease that should be treated by non-surgical means. The majority of patients, however, seem to have a more or less localized epileptogenic dysfunction without gross morphological abnormalities. Although the history sometimes suggests that a perinatal or subsequent trauma or other events might be causally related, more often no definite etiology can be established prior to surgery.

(2) The nature of epilepsy determines whether surgery is indicated or not and whether a curative or a palliative operation can be suggested:

(a) Curative (causal resections): The localization and volume of the tissue responsible for triggering the patient's habitual spontaneous seizures are known with that degree of confidence that is necessary for a surgical intervention, i.e., the "primary epileptogenic area" can be resected radically. This implies that there is *proof* for a *well delineated seizure onset zone in a brain region that can be resected without producing intolerable neurological deficits.* Epilepsies which can be treated curatively therefore express themselves with simple or complex partial seizures, with or without secondary generalization.

(b) Palliative epilepsy surgery: These procedures aim either at resection of the "secondary pacemaker focus", leaving partially or totally in place the primary epileptogenic area, which is surgically not accessible; or aim at interruption of fibre tracts; i.e. of pathways that are important for the propagation of the seizures under consideration.

An example of the first type of palliative operation is palliative amygdalohippocampectomy[11] in a patient with onset of seizures in Wernicke's speech area and with rapid propagation to the ipsilateral mediobasal limbic temporal lobe structures, which serve as a secondary pacemaker focus. An example of the second type of palliative surgery is anterior callosal section.

(3) Severity: Are the seizures themselves and/or their frequency and/or their circumstances of appearance such that the patient is subjectively and objectively severely handicaped?

(4) Drug-resistance: Is the patient's seizure tendency poorly controlled despite conscientious anticonvulsant treatment with drugs of first and second choice in monotherapy and combinations with sufficient serum levels? Or, is significant amelioration of seizures only possible at the price of unacceptable side effects and/or intoxication?

(5) A fundamental question is the *best age* at which surgical treatment should be envisaged. Today most

experts agree that surgical options should not be delayed unnecessarily, that is, *"early"* surgery is recommended. Whereas there is general agreement that a period of two years with proven drug resistance should be sufficient to recommend surgery for relatively uncomplicated seizure patients if the end of puberty has been reached, the question is more difficult in children because the seizure foci might not be stable and extratemporal resective surgery might be hampered by severe secondary problems, such as impaired rate of extremity growth, if the motor cortex is involved. There are, however, more and more centers which also advocate *"early"* surgical treatment in children with extratemporal foci.

Important arguments for advocating *"early"* surgery, even before the end of puberty, are:

(a) irreversible brain damage by the seizures themselves, (b) a frequently observed ongoing process-like behavioural deterioration, and (c) the fact that kindling-like mechanisms might occur which – from a certain point on – significantly reduce the surgical possibilities because of the establishment of independent secondary foci.

Several other factors influence the decision as to whether surgery should be contemplated. These are:

(6) Good cooperation becomes essential as soon as a presurgical evaluation with implantation of intracranial electrodes and long-term monitoring procedures is required. Good compliance is also necessary to guarantee that risky and expensive examinations are done only with therapeutic implications, and that anticonvulsant drug treatment is continued for the first 1–2

postoperative years together with a regular life style and avoidance of seizure provoking factors.

(7) Local facilities and restrictions – the existence of an efficient presurgical evaluation programme and the therapeutic surgical expertise – are, unfortunately, in reality very often *the* decisive limiting factors. The presurgical evaluation should, if necessary, be able to make use of all the modern non-invasive (MRI, SPECT, PET) and invasive diagnostic facilities, such as intracranial EEG, Amytal procedures, and electrocorticography including functional mapping.

At this point it may be justified to remark that the most advanced diagnostic efforts would be useless if the diagnostic conclusions could not be translated into adequate surgical techniques, and vice versa: the best microsurgical technique will not produce good results if the indication for the operation is not well elaborated. It is this interrelationship between diagnostic and therapeutic procedures that makes epilepsy surgery a most demanding interdisciplinary medical field.

The Preoperative Evaluation. In summary, the preoperative evaluation of candidates for surgical therapy of epilepsy aims at:

(1) Identification of those patients who will profit from surgery.

(2) Exact localization of the seizure generating structures which must be resected in order to reduce the patient's seizure tendency. It is important to emphasize that the "epileptogenic area" is not necessarily synonymous or coexistent with the "lesional zone" if present; its definition is as a positively defined area of seizure onset. Seizure onset might be focal, regional or

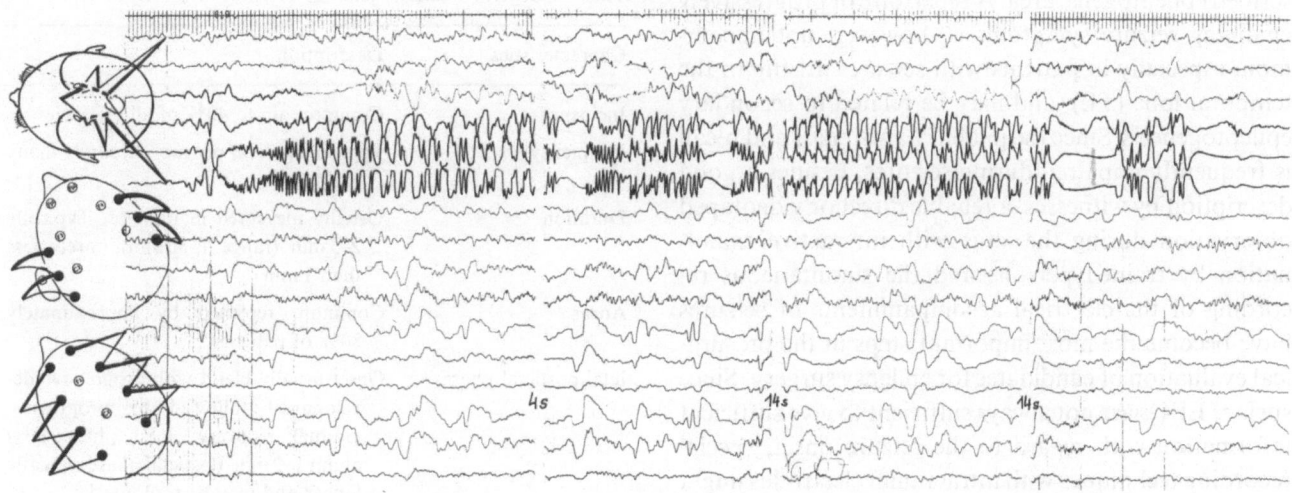

Fig. 1. Example of a focal seizure of left mesiobasal temporal origin recorded by 3-contact foramen ovale electrodes

even more diffuse. Figure 1 exemplifies a focal seizure onset recorded by foramen ovale electrodes[13].

(3) Evaluation of the potential risks of the proposed operation, i.e., the best available proof that the intended operation can be performed without producing significant neurological deficits. Important predictive information can be gained from intraoperative "functional mapping"[3] and preoperatively performed Amytal procedures.

Recently the intracarotid Sodium Amytal procedure[5] has been extended to more "selective" TL Amytal tests[2, 15, 16] with the aim of imitating temporarily the contemplated resection of mediobasal limbic structures. Three different techniques have so far been used, namely the balloon catheter technique with temporary occlusion of the internal carotid artery distal to the origin of the anterior choroidal artery, and the selective catheterization of either the anterior choroidal artery, or of the peduncular segment of the posterior cerebral artery.

Clinical Features

Weighting of Epileptic Symptoms

Knowledge of the early clinical components of a patient's seizures provides important evidence for the localization of seizure onset. Therefore, the seizure history as obtained from the patient and from witnesses should be taken carefully. This localizing evidence, when confirmed by congruous findings from other investigations, is probably the most important finding in orienting one towards consideration of surgical therapy. A stable habitual seizure pattern can be taken as a good indicator for a single and perhaps a circumscribed epileptogenic area. A repertoire of progressively changing seizure symptoms is, however, not uncommon, especially in patients with seizures arising in the temporal lobe (TL), and may be related to secondary epileptogenesis. Since the patient's awareness and recall is frequently impaired during seizures, besides a good description by witnesses, a reliable direct or videotaped observation during the ictus with interactive examination by trained persons and the simultaneous recording of the electrical accompaniments of seizures have become the most important steps in the presurgical evaluation of candidates for epilepsy surgery. Since surface EEGs are not always sufficient to give sufficient information with regard to the seizure onset, special recording techniques with intracranial electrodes might become necessary.

Clinical Characteristics of Partial Seizures

A recent proposal for classifying the epilepsies avoids the term "partial epilepsy" by including all seizure disorders with a probable localized origin in the category of "localization-related epilepsies"[1]. This scheme follows our earlier studies[6] and might be particularly appropriate when discussing seizures in the context of surgical therapy. As a consequence of our studies on the electroclinical features of the psychomotor seizure[6] we have stressed that the entire clinical episode and corresponding electrical changes need to be considered in order to develop the "gestalt" of each patient's seizure which might exhibit a characteristic "succession of symptoms". The localizational value of a single symptom or sign, when viewed out of context, remains questionable. Despite these cautious remarks we have

Table 1. *Specific Lateralizing or Localizing Characteristics of Temporal Lobe CPSs*

	Localizing	Lateralizing
Symptoms		
Elementary auditory hallucinations	Yes	No
Olfactory hallucinations	Yes	No
Vestibular hallucinations	Yes	No
Complex auditory hallucinations	Possibly	No
Complex visual hallucinations	?	No
Epigastric sensations	No	No
Signs		
Arrest reaction/motionless stare	?	No
Oroalimentary automatisms	Possibly	No
Ictal or postictal aphasia	No	?
Complex automatisms	No	No
Focal tonic/clonic motor activity	No	No

Table 2. *General Characteristics of Temporal Lobe CPSs*

Characteristics	Description
Incidence	Common, up to 40% of all epilepsies
Frequency	Typically several per month, rarely many per day
Duration	Usually measured in minutes. Typically 2–5 min (range 1–30 min), rarely less than 1 min
Auras	Common, reported by approximately 80% of patients
Ictal/postictal phases	Onset usually bland with progressive development followed by progressive gradual recovery. Ictal phase lasts about 1–2 min. Postictal phase typically longer and may be prolonged

Table 3. *Seizures of Medial Temporal (Hippocampal-Amygdalar) Origin: Clinical Characteristics*

General Characteristics	Same as Table 2
Auras	Visceral sensations (epigastric-thoracic constriction), environmental distortions (déjà vu experiences, micropsia, macropsia), emotions (fear, sadness), indescribable sensations
Consciousness	Usually minimally impaired until spread to the frontal lobe or contralateral medial temporal lobe, then maximally impaired
Early arrest reaction/ motionless stare	Occurs as the first objective sign in 30 to 50% of seizures
Autonomic changes	Common during the latter part of the seizure, but may appear as the first detectable signs
Automatisms	Initially simple, stereotyped. Oroalimentary most common. Simple stereotyped gesticular can occur initially or early in seizure. Progressive elaboration of automatic behavior. Reactive automatisms late in seizure or postictal
Motor activity	Focal tonic, general increased tone (rigidity)
Postictal	Confusion, dysphasia, partial responsiveness, gradual return
Memory	May remember aura but amnestic for "objective" seizure

Table 4. *Temporal Lobe Seizures: Clinical Findings at Seizure Onset Suggesting Origin in Lateral Temporal Neocortex*

Probable
  Elementary or simple auditory hallucinations
  Vestibular hallucinations (true vertigo)
  Focal motor or sensory manifestations (face, upper extremity)
Questionable
  Complex auditory hallucinations
  Complex visual hallucinations
  Receptive aphasia

Table 5. *Seizures of Frontal Lobe Origin: General Clinical Characteristics*

Prominent motor manifestations
  Focal tonic/clonic, including asymmetrical tonic/adversive (posturing)
  Generalized tonic/clonic
  Motor automatisms (complex bilateral, legs and arms)
Partial seizures: brief and frequent
Status epilepticus
  Convulsive
  Absence
  Focal motor
  Complex partial

Table 6. *Frontal Lobe Seizure Syndromes*

Specific regions of seizure origin
  Rolandic
  Supplementary motor region
  Cingulate gyrus
  Frontal polar (anterior frontal)
  Dorsolateral frontal
  Orbital frontal
Specific seizure types
  Considerable redundancy
  Suggestive trends
    Generalized tonic-clonic convulsions: dorsolateral frontal
    Asymmetrical tonic seizures: medial frontal (supplementary motor area)
    Absence: medial frontal (cingulate gyrus)
    Complex partial seizures: medial and orbital frontal
    Focal clonic: rolandic

Table 7. *Clinical Characteristics of Frontal Lobe Complex Partial Seizures*

Frequent attacks, often occurring in clusters of many per day
Brief seizures, lasting about 30 sec
Sudden beginning and ending with minimal postictal confusion
Prominent complex semipurposeful motor automatisms including sexual automatisms
Forced vocalization
Frequent usually non-specific warnings
Complex partial status epilepticus
Bizarre attacks that often appear hysterical
Stereotyped seizures for each patient

Table 8. *Findings That Suggest Occipital Lobe Seizure Origin*

Elementary visual hallucinations
Ictal blindness: complete or partial
Pulling or moving sensation in the eyes without detectable movement
Eye deviation, usually contralateral
Rapid forced blinking or eyelid flutter
Oculoclonic activity

described several types of complex partial seizures which we characterized by their origin and preferential spread together with a rank order of their accompanying symptoms. The following five seizure types were proposed: (1) temporal mediobasal limbic, (2) temporal pole, (3) temporal neocortical posterior, (4) opercular (insular), and (5) frontobasal-cingulate. On several occasions we have discussed the phenomenology, origin and spread of limbic[7, 8, 10, 12], neocortical temporal[14], and of frontal lobe seizures[9]. Williamson *et al.*[17] have

Table 9. *List of Evoked Symptoms and Signs with Their Frequency of Observation (n observ) in Relation to the Total of 1665 Stimulations Which Were Accompanied by Clinical Symptoms, Regardless Whether an Afterdischarge was Observed or Not*

Observed Symptoms and Signs	n observ.	% of positive stimulations
Brief movements of extremities, and/or head	202	12.1
Arrest	164	9.8
Focal twitches and cloni	142	8.6
Gestural automatisms	142	8.6
Eye movements	139	8.3
Postural changes, reactive movements, acceleration or slowing of movements, restlessness	136	8.2
Warning, unspecific	131	7.9
Respiration, change of	130	7.8
Opening or closure of eyes or mouth	128	7.7
Gastric and epigastric ascending sensation and oppression	127	7.6
Blinking	123	7.4
Oro-alimentary automatisms	115	6.9
Phonation, groaning	101	6.1
Inadequate reaction to commands, "blocked" feeling, increased reaction-time	84	5.0
Flush, pallor	83	5.0
Problems with language a/o speech	71	4.3
Dyslexia, dyscalculia	65	3.9
Cardial symptoms (tachyc., bradyc.)	57	3.4
Cephalgic sensation	57	3.4
Fear, anxiety	53	3.2
Clouding of consciousness	51	3.1
Visual (pos. + neg. sympt.)	43	2.6
Unlocalized cribbling sensation	43	2.6
Amnesia for time of stimulation	35	2.1
Vomiting, nausea	27	1.6
Smiling (pleasure?)	26	1.6
Alteration of time and space, depersonalization	26	1.6
Funny feeling	23	1.4
Staring	21	1.3
Recollection, déjà experience, dreamy state	17	1.0
Vertigo	10	0.6

tried to examine the pertinent literature describing clinical features and anatomical correlations. Their summary tables are given below (Tables 1–8).

Clinical Symptoms Provoked by Intracerebral Electrical Stimulation

To exemplify the localizational value of certain clinical signs and symptoms we briefly present some data of a comprehensive study in which the electrically evoked signs and symptoms were analyzed[4]. In 116 stereo-electroencephalographically examined patients a total of 2223 stimulations have been performed that produced either clinical and/or electrical effects. These stimulations were analyzed with regard to their anatomico-electro-clinical characteristics.

Fifty-four percent of the stimulations were in the right and 46% in the left hemisphere; 952 stimulations (43%) were accompanied by one symptom (636 stimulations) or more clinical symptoms (316 stimulations) but *without* an afterdischarge; 232 stimulations (10%) were accompanied by a *local* afterdischarge and clinical symptoms; 481 stimulations (22%) were accompanied by a *regional* afterdischarge and clinical symptoms; and 548 stimulations (25%) produced a *local* or *regional* after discharge but no clinical signs or symptoms.

Table 9 lists the observed symptoms and their relative frequency.

Figure 2 depicts six interesting symptom-groups, their frequency of appearance, and their relative distribution with regard to their evocation through stimulation of mesiobasal limbic structures (amygdala, hippocampus, parahippocampal gyrus), lateral neocortical temporal cortex, and extratemporal brain areas.

Figure 3 refers only to those 952 stimulations that were accompanied by clinical symptoms but without afterdischarges. This subgroup of electrically provoked clinical signs and symptoms therefore might be viewed as the "localisationally purest" responses. In addition, Fig. 3 illustrates a more elaborate computer-aided analysis, although the evoked signs and symptoms are displayed only according to main symptom groups. The chosen computer-display allows a comparison of different evoked signs and symptoms with regard to their localizational specificity. Since the stimulations are of course not equally distributed over the whole brain the number of total stimulations per brain area has to be taken into consideration.

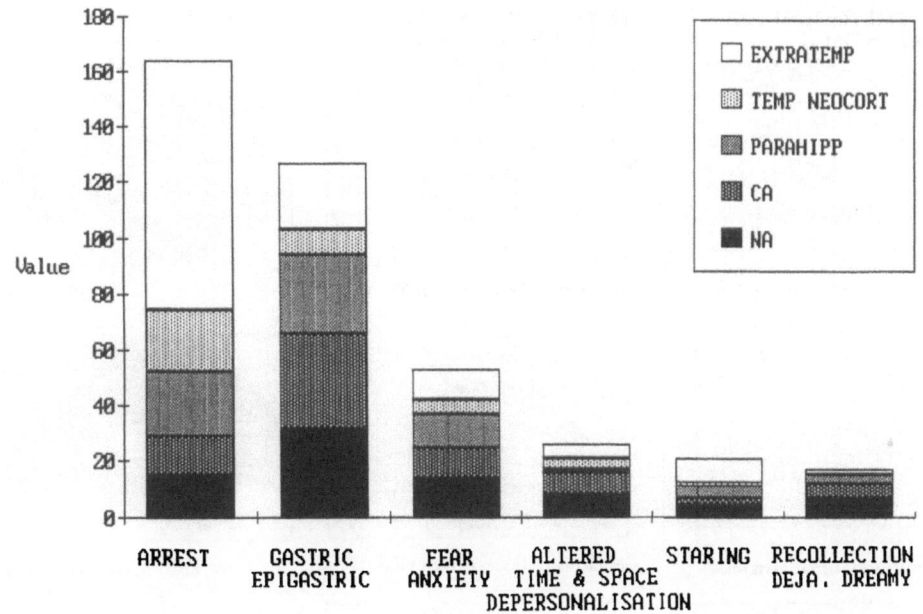

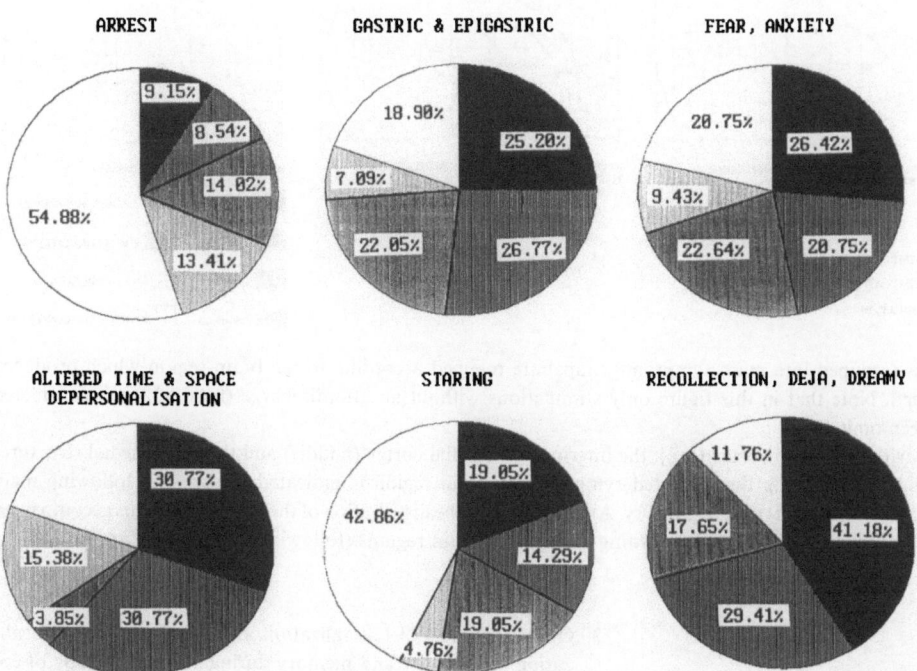

Fig. 2. Out of a total of 1665 stimulations (without or with afterdischarges) that were acompanied by clinical symptoms 408 stimulations evoked the indicated ones. For the purpose of this display the localization of the stimulations has been simplified into five brain areas: *EXTRATEMP* extratemporal; *TEMP NEOCORT* lateral temporal neocortical cortex; *PARAHIPP* parahippocampal gyrus; *CA* hippocampus; *NA* amygdala. As in Fig. 3 lateralization is not respected. i.e., both hemispheres are lumped together. To visualize better the localizational adherance of the indicated symptoms the relative distribution of brain regions which, when electrically stimulated, produced the indicated symptoms is also displayed (lower part of the figure)

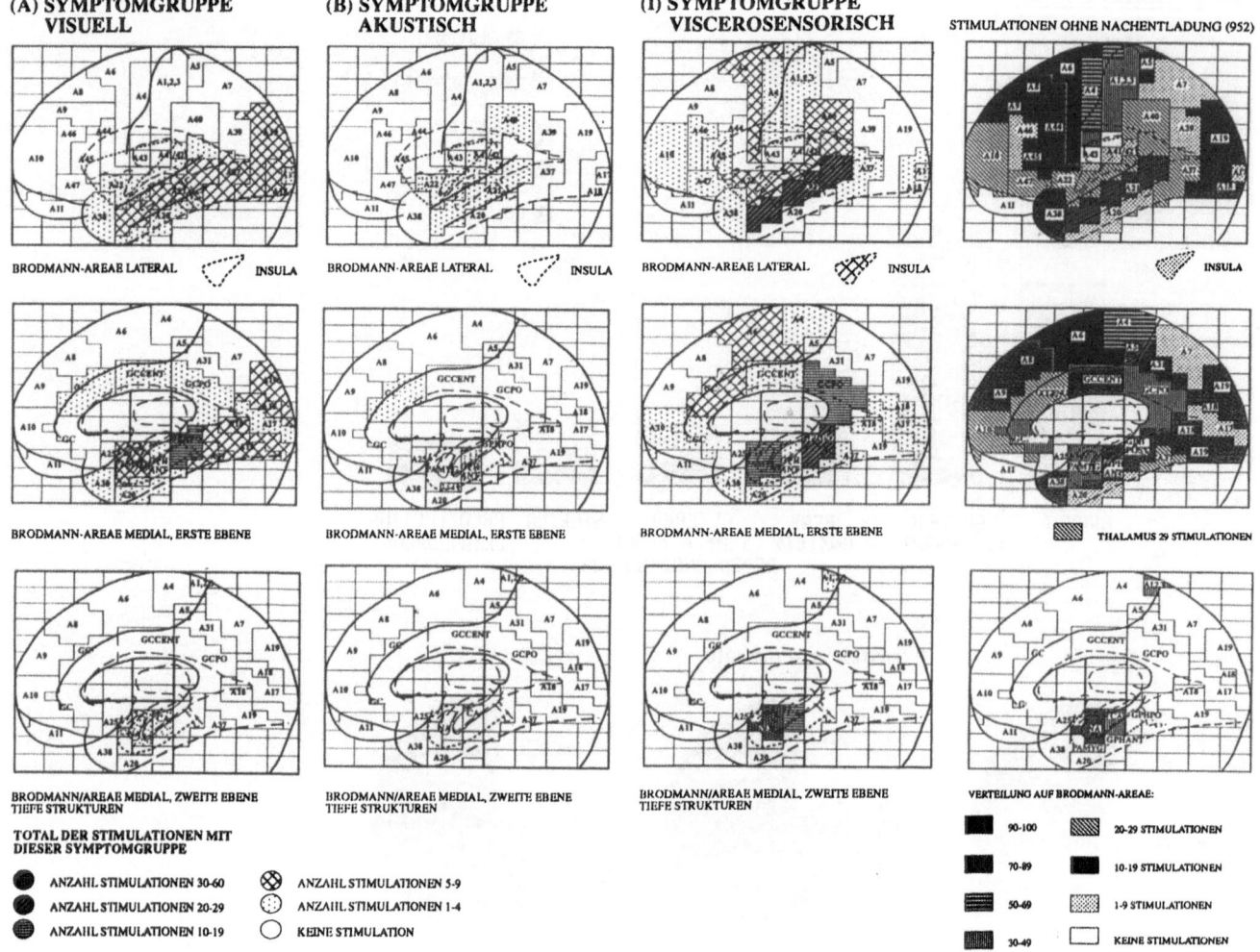

Fig. 3. A few selected evoked symptoms (grouped into main symptom-groups) are mapped according to the brain region which produced this symptom when electrically stimulated. Note that in this figure only stimulations without an afterdischarge (total 952) are considered and that lateralizational aspects have been omitted.

Each main symptom group is displayed with the lateral cortex (top), the intermediate medial cortex (middle) and the most medial structures (bottom). The number of positive stimulations evoking the indicated symptom per brain region is indicated below. The following main symptom-groups are depicted: (A) visual; (B) auditory; (I) viscero-sensory. At the very right the distribution of the 952 stimulations accompanied by clinical symptoms but without an afterdischarge is indicated according to the used brain regions (following Brodmann's areas)

Acknowledgement

Tables 1–8 are reproduced from PD Williamson, HG Wieser and AV Delgado-Escueta (ref. 17) with kind permission of the authors and Raven Press.

References

1. Dreifuss FE, Martinez-Lage M, Roger J, Seino M, Wolf P, Dam M (1985) Proposal for classification of epilepsies and epileptic syndromes. Epilepsia 26: 268–278

2. Jack CR, Nichols DA, Sharbrough FW, Marsh WR, Petersen RC (1988) Selective posterior cerebral artery Amytal test for evaluating memory function before surgery for temporal lobe seizure. Radiology 168: 787–793

3. Ojemann G (1987) Lateralization and intrahemispheric localization of language and memory during surgical therapy of epilepsy. In: Wieser HG, Elger CE (eds) Presurgical evaluation of epileptics. Springer, Berlin Heidelberg New York, pp 98–104

4. Schmid M (1990) Analyse klinischer und elektrischer Reizantworten nach zerebralen elektrischen Stimulationen. Thesis, University Zürich

5. Wada J; Rasmussen T (1960) Intracarotid injection of sodium amytal for the lateralization of cerebral speech dominance. Experimental and clinical observations. J Neurosurg 17: 266–282

6. Wieser HG (1983) Electroclinical features of the psychomotor seizure. Fischer-Butterworths, Stuttgart London

7. Wieser HG (1986) Psychomotor seizures of hippocampal-amygdalar origin. In: Pedley TA, Meldrum BS (eds) Recent advances in epileptology, vol 3. Churchill Livingstone, Edinburgh, pp 57–79

8. Wieser HG (1987) The phenomenology of limbic seizures. In: Wieser HG, Speckmann E-J, Engel J jr (eds) The epileptic focus. John Libbey, London Paris, pp 113–136

9. Wieser HG (1988) Differentiating frontal from temporal lobe seizures. Epilepsia 29 (2): 208–209

10. Wieser HG (1988) Human limbic seizures: EEG studies, origin, and spread. In: Meldrum BS, Ferrendelli JA, Wieser HG (eds) Anatomy of epileptogenesis. John Libbey, London Paris, pp 127–138

11. Wieser HG (1988) Selective amygdalo-hippocampectomy for temporal lobe epilepsy. Epilepsia 29 [Suppl 2]: S 100–S 113

12. Wieser HG, Kausel W (1987) Limbic seizures. In: Wieser HG, Elger CE (eds) Presurgical evaluation of epileptics. Springer, Berlin Heidelberg New York, pp 227–248

13. Wieser HG, Moser S (1988) Improved multipolar foramen ovale electrode recording. J Epilepsy 1: 13–22

14. Wieser HG, Müller RU (1987) Neocortical temporal seizures. In: Wieser HG, Elger CE (eds) Presurgical evaluation of epileptics. Springer, Berlin Heidelberg New York, pp 254–266

15. Wieser HG, Valavanis A, Roos A, Isler P, Renella RR (1989) "Selective" and "superselective" temporal lobe Amytal tests: I. Neuroradiological, neuroanatomical, and electrical data. In: Manelis J et al (eds) Advances in epileptology, vol 17. Raven Press, New York, pp 20–27

16. Wieser HG, Landis T, Regard M, Schiess R (1989) "Selective" and "superselective" temporal lobe Amytal tests: II. Neuropsychological test procedure and results. In: Manelis J et al (eds) Advances in epileptology, vol 17. Raven Press, New York, pp 20–33

17. Williamson PD, Wieser HG, Delgado-Escueta V (1987) Clinical characteristics of partial seizures. In: Engel J jr (ed) Surgical treatment of the epilepsies. Raven Press, New York, pp 101–120

Correspondence: Prof. Dr. med Heinz Gregor Wieser, Neurologische Klinik, Universitätsspital, Frauenklinikstrasse 26, CH-8091 Zürich, Switzerland.

Editor's Note

For a detailed account of depth electrode implantation see Munari C (1987) Depth electrode implantation at Hopitale Sainte Anne, Paris. In: Engel J jr (ed) Surgical treatment of the epilepsies. Raven Press, New York, pp 583–588.

Acta Neurochirurgica, Suppl. 50, 72–75 (1990)

Selection Criteria for Epilepsy Surgery Psychometric Evaluation

C. E. Polkey

Neurosurgical Unit, Maudsley Hospital, London, U.K.

Summary

The principles underlying psychometric evaluation of the patients for epilepsy surgery are reviewed together with the part which such studies can play in defining which surgical procedure is appropriate to a particular patient, and in revealing abnormalities of cerebral function and organisation that may be present.

Keywords: Epilepsy; operative treatment; psychometric evaluation; cognitive functions; recognition memory; prognosis.

Introduction

This review is not written as a practical guide to the psychometric evaluation of patients prior to epilepsy surgery since this is the specialised task of an appropriately trained neuropsychologist. It is rather an introduction to the principles involved and to the part which such studies can play, both in deciding which surgical procedure is appropriate to a particular patient, and in disclosing anomalies of cerebral function and organisation which often occur in such patients.

The majority of successful operations for epilepsy consist of the resection of cerebral tissue containing a significant pathological lesion or a clear electrical focus as a basis for the patient's epilepsy. Operations based upon altering the functional activity of the brain such as stereotactically placed lesions or callosotomy are less successful[19]. In identifying such candidates it is necessary to show both from the clinical history, and from the results of neurophysiological, neuroradiological and neuropsychological investigations, that the seizures originate from a single area of the brain which can be resected without gross adverse effects. Because 60–70% of cerebral resections for epilepsy are from the temporal lobes, their cognitive function, which is mainly memory function, is of great interest.

However, although the collection of neuropsychological data from both adults and children is relatively easy, the interpretation of the results is difficult. There are a number of reasons for these difficulties which may be summarised as follows.

First, especially in assessing memory function, the data collected is relatively crude and often examines only one aspect of memory. Secondly, as a result of the original brain insult, which is also responsible for the epilepsy, there may have been reorganisation of cognitive function. Such a reorganisation, most easily revealed as unusual cerebral dominance for speech, may make the interpretation of the results of neuropsychological assessment difficult. Lastly, it is known that the electrical dysfunction accompanying chronic epilepsy, and representing neuronal malfunction, may also give rise to cognitive impairment which will resolve if this activity can be abolished or controlled.

Testing

Neuropsychological assessment of these patients is undertaken in all centres offering surgical treatment for drug-resistant epilepsy. It consists of a relatively standard battery of tests including measurement of Full Scale IQ, Verbal IQ and some form of Spatial or Performance IQ. In addition specific tests of frontal lobe function such as the Wisconsin Card sorting test, and tests which quantify recent and delayed memory, both verbal and non-verbal are given to the patient. Suitable normative data are available for these tests in both adults and children.

Cerebral dominance for speech and the distribution of memory are clearly of great importance and may be assessed using the Wada intra-carotid amytal test[22] or one of its variants. Originally intended as a test of speech representation it was later modified as a test for global amnesia following temporal lobe surgery[10]. In the commonly used version of this test one or other cerebral hemisphere is briefly anaesthetised. A catheter is introduced into the internal carotid under local anaesthesia and a suitable dose of sodium amytal, usually 75–150 mgs, is injected. The efficacy of the injection is judged by the degree of hemiplegia and changes in the scalp EEG. There are minor variations in the conduct of the test, the materials used for testing memory vary between centres, and some centres examine each hemisphere on separate days, but on the whole the results of the test are reliable[17].

However there are theoretical objections to the test in its traditional form. More of the hemisphere is anaesthetised than is necessary for the memory testing and the aphasia and behavioural changes which can occur may make the administration of the test material and the interpretation of the results of the testing difficult. To overcome these problems two methods of selective injection have been introduced recently. In one from Zurich, the anterior choroidal artery is selectively catheterised and then injected. In the other, from the Mayo clinic,[4] selective injection is made into the posterior cerebral circulation. Both of these methods are claimed to be accurate predictors of the results of surgery, although how they take account of the fact that the blood supply of the hippocampus is divided between the anterior and posterior cerebral circulations and this variation probably differs from one individual to another is not stated.

Basis of Memory and Other Cognitive Functions

The basis of normal memory is complex and clinical evidence and the results of animal experiments throw light upon it in a very crude way. Until recently most of the tests of memory both in experimental animals and in the clinical situation were of recognition memory. In clinical material distinction could be made between antegrade and retrograde memory. It has been shown in animal experiments that, as well as the temporal lobes, other structures including the thalamus contributed significantly to memory and that memory deficits were related to the extent of the lesions produced and that in general, if lesions were bilateral, then the more extensive the lesion the more extensive was the memory loss[21]. It is also known that in patients a similar situation occurs. Severe damage caused by encephalitis as described for their patient DRB by Squire

and Zola-Morgan[21], or by surgeons as described in the case of the patient HM[18], results in a severe retrograde and antegrade amnesia. In another RB, also described by Squire and Zola-Morgan[21], the damage after an ischaemic insult, although bilateral, was restricted to the CA 1 portions of the hippocampus and there was no retrograde amnesia. Both experience with HM, and more recent work has demonstrated that it is possible to separate short-term and long-term memory. It has also been shown in animal experiments that if small removals of the hippocampus or amygdala are carried out bilaterally they can substitute for each other when recognition memory is tested but if association memory is tested then after amygdalar removal a significant deficit is seen[11].

The effect of cerebral reorganisation due to preexisting disease also has to be considered. A key paper is that published by Rasmussen and Milner in which they describe the effect of early left-brain injury on speech representation at maturity[16]. They showed that although there was a strong tendency for speech to remain in the left hemisphere, if the cerebral insult occurred before the age of six years, then there was a tendency for speech to transfer to the right hemisphere whereas after that age any recovery of speech function was due to intra-hemispheric reorganisation.

Finally it has been shown that interictal spike discharges in epileptics could be related to cognitive function[26], and furthermore that transient cognitive impairment accompanying subclinical discharges could be related to the site of those discharges[1]. It is also known that the progressive abnormal electrical activity induced in the normal hemisphere by unilateral disease in the opposite hemisphere can be severe enough to produce dementia, this being one of the indications for hemispherectomy suggested by Krynauw[6]. When such an operation is successful there can be a considerable increase in the Full Scale IQ score[15,2].

Interpretation of Cognitive Functioning

In this section some of the more subtle effects of surgery upon cognitive function will be discussed. Many years ago Milner and her colleagues demonstrated that material-specific cognitive deficits would follow unilateral temporal lobetomy[8,9]. In a comprehensive review of the Montreal material Benzgon et al. looked at those factors which made for success in temporal lobectomy[3]. They showed that in patients with a good outcome the neuropsychological deficits were found only in two tests of temporal lobe function, and furthermore they were well lateralised. A number of studies have shown that patients with a low overall IQ score, presumably indicative of diffuse brain damage, have poorer results from temporal lobe surgery[23, 17]. However it has also been shown both by ourselves and the Los Angeles group that the effect of operation upon cognitive function is related to the pre-operative function of the resected temporal lobe. In a study of 59 patients operated upon at the Maudsley Hospital, Powell et al. found that the younger, intellectually less able patients tended to be less affected by the resection[14].

It was hoped to correlate neuropsychological findings with the pathological findings in the resected temporal tissue. In a subsequent study of 40 patients, 60% of whom had mesial temporal sclerosis in the resected specimen, 15% non-specific changes and 25% other lesions, we were unable to demonstrate any direct correlation between pathology and cognitive deficit[7]. However there was a difference in verbal memory between patients with MTS and those with other lesions. In those with a lesion other than mesial temporal sclerosis in the resected temporal lobe, then if it was the dominant temporal lobe then verbal memory was poorer than average both pre-operatively and post-operatively, and if it was the non-dominant temporal lobe the verbal memory was above average. The changes for non-verbal function in the same patients were non-significant. Similar changes have been described in the Montreal material[5].

Now mesial temporal sclerosis is a lesion which is usually acquired in childhood between the age of one to two years, at a time when the cerebral hemispheres have a potential for re-organisation. This was to some extent borne out in a study of a small group of 27 patients subjected to carotid amytal tests, 16 of whom subsequently underwent surgery[13]. A high proportion of these patients were presumed to have mesial temporal sclerosis as the causative lesion in their temporal lobe and this was subsequently demonstrated at surgery in some of them.

Language was found to be in the left hemisphere in 63% of these patients. If this is compared with the data from the study by Rasmussen and Milner[16] then it is clear that some of our patients must have transferred their dominance from the left hemisphere since left hemisphere location of language was found in 83% of the undamaged patients in Rasmussen and Milner's study. This was reflected in the distribution of memory function. Whereas one would expect in a normal population to find memory function equally divided between the hemispheres, only 11% of our patients showed this arrangement. In 44% of the patients all memory appeared to be in one hemisphere and contralateral to the lesion and in a further 7% almost so. In 22% neither hemisphere appeared to be able to sustain normal memory alone suggesting severe damage with a bilateral element, and in 15% the majority of the memory seemed to be in the lesioned hemisphere.

Although it could be surmised that removal of the amygdala and hippocampus would have cognitive sequelae the accurate prediction of these is very difficult. Corsi showed that only the removal of a large portion of the hippocampus in the non-dominant hemisphere resulted in a detectable impairment of spatial memory[17]. By contrast Ojemann and Dodrill[12] have shown that by limiting the amount of neocortical removal at temporal lobectomy in the dominant hemisphere they can minimise the verbal memory deficit which may follow such operations. However, Spencer and colleagues[20] have reported that if the neocortical removal at temporal lobectomy is limited to 4.5 cms from the pole then an extensive hippocampal removal can be carried out with no cognitive penalty. Finally Wieser, in reporting the results of unilateral selective amygdalo-hippocampectomy, notes that there is no change in cognitive function in the majority of these patients[24]. It has been impossible to demonstrate a consistent relationship between the size of the hippocampal resection and specific cognitive deficits although Jones-Gotman described some tasks which are impaired following large hippocampal resections but not small ones and this seems to be related to the way in which the material is processed[5].

None of these studies take account of possible dynamic factors affecting cognitive function. As mentioned previously it has been clearly shown that cognitive function can be affected on a shorter time scale by epileptic discharges. Wieser and Landis have reported four patients with unilateral limbic status epilepticus, as seen on stereo-EEG recordings, in whom they examined their cognitive function prior to operation[25]. In one of these cases they describe how left (dominant) hemisphere cognitive function was affected during localised limbic discharges within that temporal lobe recovering to some extent in between. In addition, in the same patient the right (non-dominant) cognitive function was enhanced and during the left limbic discharges this enhancement was increased. They suggest that in this patient there was a protective inhibition which protects the opposite hemisphere from the effects of the electrical discharge. After left amygdala-hippocampectomy the patient's seizures ceased and her neuropsychological profile returned to normal.

References

1. Aarts JHP, Binnie CD, Smith AM, Wilkings AJ (1984) Selective cognitive impairment during focal and generalised epileptiform EEG activity. Brain 107: 293–308

2. Beardsworth ED, Adams CBT (1988) Modified hemispherectomy for epilepsy: Early results in 10 cases. Br J Neurosurg 2: 73–84

3. Benzgon ARA, Rasmussen T, Gloor P, Dassault J, Stephens M (1968) Prognostic factors in the surgical treatment of temporal lobe epileptics. Neurology 18: 717–731

4. Jack CR, Nichols DA, Sharbrough FW, Marsh WR, Petersen RC (1988) Selective posterior cerebral amytal test for evaluating memory function before surgery for temporal lobe seizure. Radiology 168: 787–793

5. Jones-Gotman M (1987) Commentary: Psychological evaluation – testing hippocampal function. In: Engel J (ed) Surgical Treatment of the Epilepsies. Raven Press, New York, pp 203–211

6. Krynauw RA (1950) Infantile hemiplegia treated by removing one cerebral hemisphere. J Neurol Neurosurg Psychiatry 13: 243–267

7. McMillan TM, Powell GE, Janota I, Polkey CE (1987) Relationships between neuropathology and cognitive functioning in temporal lobectomy patients. J Neurol Neurosurg Psychiatry 50: 167–176

8. Milner B (1954) Intellectual function of the temporal lobes. Psychol Bull 51: 42–62

9. Milner B (1958) Psychological defects produced by temporal-lobe excision. Res Publ Assoc Res Nerv Ment Dis 36: 244–257

10. Milner B, Branch C, Rasmussen T (1962) Study of short-term memory after the intracarotid injection of sodium Amytal. Trans Am Neurol Assoc 87: 224–226

11. Murray EA, Mishkin M (1985) Amygdalectomy impairs cross-modal association in monkeys. Science 228: 604–606

12. Ojemann GA, Dodrill CB (1985) Verbal memory deficits after left temporal lobectomy for epilepsy: Mechanism and intra-operative prediction. J Neurosurg 62: 101–107

13. Powell GE, Polkey CE, Canavan AGM (1987) Lateralisation of memory function in epileptic patients by use of the sodium amytal (WADA) technique. J Neurol Neurosurg Psychiatry 50: 665–672

14. Powell GE, Polkey CE, McMillan TM (1985) The new Maudsley series of temporal lobectomy I: Short term cognitive effects. Br J Clin Psychol 24: 109–124

15. Rasmussen T (1987) Cortical resection for multilobe epileptogenic lesions. In: Wieser HG, Elger DE (eds) Presurgical evaluation of epileptics. Springer, Berlin Heidelberg New York, pp 344–351

16. Rasmussen T, Milner B (1977) The role of early left-brain injury in determining lateralization of cerebral speech functions. Ann N Y Acad Sci 299: 355–369

17. Rausch R (1987) Psychological evaluation. In: Engel J (ed) Surgical treatment of the epilepsies. Raven Press, New York, pp 181–195

18. Scovile WB, Milner B (1957) Loss of recent memory after bilateral hippocampal lesions. J Neurol Neurosurg Psychiatry 20: 11–21

19. Spencer DD (1987) Postscript: Should there be a surgical treatment of choice and if so how should it be determined? In: Engel J (ed) Surgical treatment of the epilepsies. Raven Press, New York, pp 477–484

20. Spencer DD, Spencer SS, Mattson RH, Williamson PD, Novelly RA (1984) Access to the posterior medial temporal lobe structures in the surgical treatment of epilepsy. Neurosurgery 15: 667–671

21. Squire L, Zola-Morgan S (1988) Memory: brain systems and behavior. TINS 11: 170–175

22. Wada J (1949) A new method for the determination of the side of cerebral speech dominance. A preliminary report on the intracarotid injection of sodium amytal in man. Med Biol 14: 221–222

23. Wannamaker BB, Matthew CG (1976) Prognostic implications of neuropsychological test performance for surgical treatment of epilepsy. J Nerv Ment Dis 163: 29–34

24. Wieser HG (1985) Selective amygdala-hippocampectomy: indications, investigative technique and results. In: Symon L et al (ed) Advances and technical standards in neurosurgery, Vol 13. Springer, Wien New York, pp 39–133

25. Wieser HG, Hailemariam S, Regard M, Landis T (1985) Unilateral limbic epileptic status acitivity: Stereo EEG, behavioral and cognitive data. Epilepsia 26: 19–29

26. Wilkus RJ, Dodrill CB (1976) Neuropsychological correlates of the electroencephalogram in epileptics: I Topographic distribution and average rate of epiletiform activity. Epilepsia 17: 89–100

Correspondence: C. E. Polkey, M.D., F. R. C. S., Neurosurgical Unit, Maudsley Hospital, De Crespigny Park, London, SE5 8AZ, U.K.

Acta Neurochirurgica, Suppl. 50, 76–79 (1990)

Neuroradiology Including Magnetic Resonance

P. Anslow

Department of Neuroradiology, Radcliffe Infirmary, Oxford, U.K.

Summary

The radiological assessment of drug resistant temporal lobe epilepsy has changed dramatically over the last few years. New CT scan techniques and the increasing application of MRI have meant that ever more subtle lesions can now be regularly identified. The major challenge now lies in the accurate identification of atrophic lesions such as Ammon's horn sclerosis.

Keywords: Computerised tomography; magnetic resonance imaging; temporal lobe epilepsy.

Introduction

Carefully performed, high resolution CT and MRI are the cornerstones of modern radiological investigation and pre-operative assessment of patients with drug resistant temporal lobe epilepsy (DRTLE).

Both CT and MRI are very sensitive in detecting focal abnormalities, although neither are very specific concerning the nature of the abnormalities demonstrated. However, both techniques are relatively insensitive when assessing atrophic pathologies (Ammon's horn sclerosis).

Major degrees of atrophy are easily and confidently detected, but as the lesions become more subtle, confidence levels fall. Coronal plane studies (after the administration of intrathecal contrast in CT but much more easily obtained with MRI) and careful analysis of the contour of the temporal lobe may offer the best chance of assessment of subtle degrees of atrophy.

Methods and Material

Computerised Tomography

Routine investigation of patients should always consist of:

1) A conventional CT brain scan examining the whole of the cranium.

2) A scan performed in the plane of the temporal lobe after the administration of intravenous contrast.

When investigating any patient with epilepsy, the whole of the brain must first be examined to provide a basis for subsequent more specific and more highly localised studies.

The technique for examining the brain in the plane of the temporal lobe has been described elsewhere. A "topogram" or "scout" image is first obtained and a plane with approximately 20 degrees of caudal tilt selected which corresponds to the axis of the temporal horns of the lateral ventricles. Thin (4 or 5 mm) scans are then obtained starting from the floor of the middle fossa and ending above the sylvian fissure. Intravenous contrast enhancement must be given to identify small vascular abnormalities and neoplasms.

Focal lesions will be seen easily using these techniques, but if the scan is initially interpreted as normal, great care must be taken to ensure that a small focal lesion is not missed and specific attention should be directed towards the following radiographic features:

1) Asymmetry of the temporal horns.

In a normal patient, the temporal horns will be seen as a thin line of CSF density, with a small expansion distally. The tip of the temporal horn itself will only be seen on one or two scan slices. The two horns should be absolutely symmetrical. Any asymmetry raises the possibility of a mass lesion or atrophic process.

2) Distances between the temporal horn and the medial surface of the temporal lobe.

Assuming that no obvious parenchymal abnormality has been revealed, measurement of the thickness of the medial temporal structures may reveal an occult mass lesion. It is usually easy to accurately identify the medial aspect of the temporal lobe against the suprasellar cistern and cerebral peduncles. If difficulty is experienced, intrathecal contrast enhancement is always helpful.

3) Asymmetry of the suprasellar cistern about the midline.

Whilst the cistern is variable in shape, it should always be absolutely symmetrical about the midline. Any distortion demonstrated may again be a sign of a mass lesion or atrophic process.

It is important to stress that if no parenchymal lesion is visible, the above three signs may be the only evidence of a structural lesion. Any mass lesion will usually affect more than one of these and confidence in the diagnosis of a mass lesion rises with each positive sign. If care has been taken in the performance and interpretation of the CT scan, it is highly unlikely that a focal temporal lobe lesion will be overlooked or that any major atrophic pathology will not be identified. More subtle atrophic processes however will require further radiological assessment.

Intrathecal Enhanced CT

The introduction of contrast into the subarachnoid space delineates the contour of the medial temporal lobe structures more precisely, and permits greater accuracy in measurement. Further information can be obtained if scans are then obtained in the coronal plane. The relationship of the temporal horn to the medial temporal structures can be more readily appreciated in this plane, and measurements can be obtained of both the medial temporal and also of the inferior temporal structures. The "overlay" technique described elsewhere in this account can also be applied to these studies.

Intrathecal contrast enhanced CT is an invasive examination. Some patients will suffer lumbar puncture headache and a small minority may experience an increase in seizure activity. The examination may prove technically unsatisfactory if artefacts from tooth fillings degrade important structures. These limitations and complications limit the usefulness of the technique and if modern high resolution MRI scans are available, the examination is not indicated.

Magnetic Resonance Imaging

Multislice techniques utilised in modern machines permit the acquisition of both axial and coronal T 1 and T 2 weighted images in a short time (typically 10 minutes for each plane in 1.5 Tesla machines). Images demonstrate temporal lobe anatomy extremely well, and the grey and white matter can be clearly delineated.

The fine structure of the hippocampus remains beyond the resolving power of machines in routine use.

The usual scan plane chosen for MR imaging is quite different to that performed in CT. The gantry cannot be angled as in CT and although any scan plane can theoretically be chosen, deviations from the major axes of the machine will result in some loss of image quality. If the patient's head is slightly extended, then the scan plane approximates to that of the temporal lobes and the advantages described for CT scans are again achievable. Focal parenchymal abnormalities will be demonstrated easily and the vast amount of experience available would suggest that MRI is significantly more sensitive than CT in detecting parenchymal lesions. Contrast enhancement can be obtained with gadolinium DIPA, but this is an extremely expensive agent and its role in the routine investigation of patients is unproven. As with CT, the major problem with MRI is in the identification of subtle atrophic pathologies. As the lesions become more subtle, the degree of confidence with which they can be diagnosed decreases.

The same interpretive skills described for the assessment of CT scans should be directed to the MRI images. Focal lesions can usually be more clearly seen, but more importantly in the context of epilepsy surgery their precise anatomical extent can usually be more clearly ascertained. This is partly due to the greater contrast sensitivity of MRI but also due to the use of multiple imaging planes, including sagittal scans. It is usually possible to identify precisely which gyri are affected, and the exact relationship of any neoplasm to adjacent important structures.

Radiological Assessment of Atrophic Pathologies-Coronal Scans and the Assessment of Subtle Atrophic Changes

Coronal plane images obtained from intrathecal enhanced CT or MRI studies offer a possibility for the assessment of subtle atrophic pathologies. Measurement of the absolute temporal lobe volume has proven impossible because of the lack of any definable point of the posterior limit of the temporal lobe on axial or coronal scans. A major problem arising when assessing the temporal lobe for any focal volume loss is the presence of any degree of obliquity in positioning of the patient in the scanner. The question posed then is whether the difference in size seen on any individual scan slice reflects true differences in the temporal lobe or is merely a feature of positioning.

In an attempt to resolve this we have recently been

experimenting with an "overlay" method. The contour of the temporal lobes on a series of coronal images is traced onto a transparency. The temporal horn is also included in the tracing to provide a standard reference point. The outlines of one temporal lobe are then inverted laterally and superimposed upon the other side. If obliquity alone is responsible for the apparent asymmetry then it should be possible to fit one outline on another at some point. If this cannot be achieved, then true asymmetry must exist.

Results

In the period 1982–1990 almost 100 temporal lobotomies have been carried out in Oxford for intractable epilepsy. During this period, a number of technical advances have been made, notably the use of a high resolution CT scanner, the routine use of intrathecal contrast (since 1986), special scan planes and more recently MRI. From an imaging viewpoint therefore the investigational protocol has changed and developed over this period. Great attention has been given to the identification of focal pathology, whenever possible.

Focal Lesions

Smaller lesions were identified during the second half of this period than during the first half. On no occasion, when focal pathology was predicted, was no abnormality discovered on subsequent pathological examination of the specimen, although the prediction of the precise nature of the pathological process was occasionally incorrect. On only two occasions was a glioma detected on examination of the pathological specimen which was not demonstrated radiologically. Both ot these patients were investigated before intrathecal studies and MRI were introduced. Removal of the focal lesion led in most cases to a complete or major reduction in seizure frequency (see Table 1).

Table 1. *% Seizure-Free Patients After Temporal Lobectomy – 2-Year Follow-up*

	"Ammonshorn" group	"Focal pathology" group
Overall	63%	68%
Good Prognosis Group	85%	78%

Atrophic Lesions

When an obvious atrophic process or significant asymmetry, not caused by a mass lesion, was demonstrated on CT, subsequent pathological demonstration always revealed an 'atrophic' pathology. In these cases the pathological changes were always severe.

In those cases when no abnormalities was discovered, or when the changes were so subtle as to be of doubtful significance, the commonest pathology was Ammon's horn sclerosis or some other 'atrophic' process. Where radiology was unhelpful in predicting the side of pathology, reliance was made on other criteria: clinical, psychological or electroencephalographic. Scans in the coronal plane, enabled by magnetic resonance, may well prove helpful in identifying these subtle lesions, and abnormal signal from gliotic brain may well have a predictive value.

Discussion

If MRI is available, it is quite clear that its supreme contrast sensitivity make it the imaging method of choice in the assessment of patients with DRTLE for the identification of focal parenchymal abnormalities. Careful CT studies however can demonstrate very small abnormalities and in certain cases, especially when the lesion is heavily calcified and where the soft tissue component is small, CT can be easier to interpret. The two studies are therefore complimentary and not exclusive.

It is also clear that MRI may not be generally available at present and that many machines in use today are of an older generation which do not have the very high spatial resolution required to identify very subtle degrees of tissue loss and atrophy. Even with modern high field equipment very thin scan slices (easily achieved by CT) cannot be obtained, and MR scan sequence protocols normally require a small gap between each slice further limiting the absolute spatial resolution which the system can achieve. These limitations in spatial resolution are usually greatly outweighed by supreme contrast resolution and only become significant when the question posed is one of the identification of subtle structural abnormalities, specifically, the fine atrophic changes of Ammon's horn sclerosis.

The "overlay" method described above seems quite convincing, but it remains to be seen whether the changes demonstrated do actually reflect pathologically demonstrable cerebral tissue loss and at present it seems that it is an interesting avenue for research, which has yet to be proven.

References

1. Aaron J, New P, Strand R, Beaulie UP, Elmden K, Brady T (1984) NMR imaging in temporal lobe epilepsy due to gliomas. J Comput Assist Tomogr 8: 608–613
2. Abou-Khalil B, Sackellares J, Latack J, Vanderznt C (1984) Magnetic resonance in refractory epilepsy. Epilepsia 25: 650
3. Bauer G, Mayr U, Pallua A (1980) Computerised axial tomography in chronic partial epilepsies. Epilepsia 21: 227–233
4. Blom R, Vinuela F, Fox A, Blume W, Girvin J, Kaufman J (1984) Computed tomography in temporal lobe epilepsy. J Comput Assist Tomogr 8: 401–405
5. Bolender N, Wyler A (1982) CT measurements of mesial temporal lobe herniation in epileptics. Epilepsia 23: 409–416
6. Engel J, Driver M, Falconer M (1975) Electrophysiological correlates of pathology and surgical results in temporal lobe epilepsy. Brain 98: 129–156
7. Falconer M, Serafetinides E (1963) A follow-up study of surgery in temporal lobe epilpsy. J Neurol Neurosurg Psychiatry 26: 154–165
8. Gammal T, Adams R, King D, So E, Gallagher B (1987) Modified CT techniques in the evaluation of temporal lobe epilepsy prior to surgery. AJNR 8: 131–134
9. Heinz E, Heinz T, Radtke R, Darwin R, Drayer B, Fram E, Djang W (1988) Efficacy of MR vs CT in epilepsy. AJNR 9: 1123–1128
10. Laster D, Penry J, Moody D et al (1985) Chronic seizure diseases: contribution of MR imaging when CT is normal. AJNR 6: 177–180
11. Latack J, Abou-Khalil B, Siegel G, Sackellares J et al (1986) Patients with partial seizures: evaluation by MR, CT, and PET imaging. Radiology 159: 159–163
12. Molyneux A, Anslow P, Easterbrook P, Oxbury J, Adams C, Hughes J (1986) The radiologic investigation of temporal lobe epilepsy. Acta Radiol [Suppl] 369: 400–402
13. Radtke R, McNamara J, Lewis D, Heinz E (1986) Usefulness of MRI in pre-surgical evaluation of intractable complex partial seizures. Epilepsia 27: 612
14. Turner D, Wyler A (1981) Temporal lobectomy for epilepsy: mesial temporal herniation as an operative and prognostic finding. Epilepsia 22: 623–629
15. Wyler A, Bolender N (1983) Preoperative CT diagnosis of mesial temporal sclerosis for surgical treatment of epilepsy. Ann Neurol 13: 59–64

Correspondence: P. Anslow, M.D., X-Ray Department, Radcliffe Infirmary, Oxford OX2 6HE, U.K.

Acta Neurochirurgica, Suppl. 50, 80–83 (1990)
© by Springer-Verlag 1990

SPECT in the Presurgical Evaluation of Patients with Temporal Lobe Epilepsy – a Preliminary Report

A. R. Andersen[1], **G. Waldemar**[1], **M. Dam**[2], **A. Fuglsang-Frederiksen**[3], **M. Herning**[4,5], and **Chr. Kruse-Larsen**[6]

[1] University Clinic of Neurology, Rigshospitalet, University Clinic of [2] Neurology, of [3] Clinical Neurophysiology, of [4] Magnetic Resonance Imaging, of [5] Radiology and [6] Neurosurgery, Hvidovre Hospital, Copenhagen, Denmark

Summary

Twenty-eight patients with drug resistant temporal lobe epilepsy (DRTLE) were studied using single photon emission computed tomography (SPECT) using xenon-133 inhalation and Tc99 m-d, 1-HMPAO with TOMOMATIC 64 as part of a presurgical evaluation programme. The visually evaluated flow-images were studied after blinding and the results subsequently compared to the EEG, MRI and CT scanning studies. In 24 patients a significant low flow region was seen in one temporal lobe. The SPECT result corresponded to the EEG findings in all but 6 patients. In 2 of these patients no side localization was indicated by EEG, while in four patients the EEG suggested that the opposite side was epileptogenic. Ictal SPECT and MRI/CT agreed with the resting SPECT study in three patients, while one patient has remained undiagnosed with respect to side-localization. In 14 patients discordance between SPECT and the CT scan was observed, but in 11 of these patients the SPECT study correlated with the other focal diagnostic tools. In 11 of the 21 patients studied by MRI, the results corresponded to SPECT; in 7 of the 10 which did not correspond, ictal studies of EEG and SPECT defined the side, in four of these 7 patients. Using the neuroimaging tools in concert 16 patients have been selected for surgery. All patients have benefited from surgery.

These preliminary results correspond favourably with earlier studies comparing SPECT and PET with CT, MR and EEG.

Keywords: SPECT; partial epilepsy; Technetium-99 m-HMPAO; Xenon-133; cerebral circulation.

Introduction

Single photon emission computer tomography (SPECT) of regional cerebral blood flow (rCBF) changes is measuring a functional parameter, while magnetic resonance imaging (MRI) yields structural (anatomical) information. SPECT of rCBF has been valuable in the description of major cerebral disorders: stroke[22], dementia[19] and classic migraine[1]. In the last decade the SPECT studies of rCBF changes in patients suffering from epilepsy have also been of increasing importance.

It has been known for many years that both generalized and focal epileptic seizures are associated with an increase of blood flow and of oxidative metabolism[15,16,3]. Using the intra-arterial xenon-133 clearance technique with stationary gamma – detectors (not SPECT), Ingvar clearly demonstrated the ictal flow increase in patients with partial epilepsy and he reported that, in the interictal state, the flow in the focal area was decreased below the normal level[7].

Several studies using positron emission tomography (PET) and tracers of brain metabolism or blood flow extended these early observations to larger series of patients[13,5,9,21]. Engel and associates[5] used PET imaging as an aid to localize the affected side in drug resistant cases of temporal lobe epilepsy (DRTLE) that were sufficiently severe to be candidates for temporal lobectomy[13,5].

Similar studies based on SPECT have now been published from several groups reporting both ictal and interictal changes of CBF distribution[2,6,12,17,18,20].

SPECT Methodology

The basic instrumentation is a SPECT, also simply called a rotating gamma camera. Essential in such an instrument is the perforated lead shield, *the collimator*, positioned between the head of the patient and the detectors. It determines the direction of the photons (gamma-rays) seen by the camera. The camera permits the recording of a series of evenly spaced views ("projections") of the head. This is accomplished by having the whole camera rotate around the head or by having the collimator alone rotating. The data sampling lasts typically up to 20 to 30 minutes when using tracers that are well retained in the brain (see below) and 4 to 5 minutes with freely diffusible tracers such as xenon-133. The projection data are reconstructed transaxially so that a set of transverse images is obtained basically of the same type as with CT scanning. Stacking these images, coronal and sagittal planes or sections parallel to the temporal lobes can also be reconstructed.

Conventional single-headed low speed SPECT instruments yield a resolution of about 15–20 mm (FWHM). Brain dedicated SPECT systems have a spatial resolution of 9 to 10 mm in the plane and axially using Tc99 m. Some SPECT devices are capable of recording rapid (dynamic) tomograms albeit with lesser resolution. It should be noted that the spatial resolution now routinely obtainable for CBF with SPECT is comparable to that obtained for CBF by PET.

The isotopes

Xenon-133 is a rather weak photon emitter (0.81 keV). As it arrives in the brain and washes out quite rapidly[4, 10] a fast, highly sensitive SPECT system is needed for scanning.

Tracers for CBF labelled with Tc99 m or Iodine-123 are also available. These isotopes have physical properties more suitable for SPECT. The labelled molecules have been designed to be retained in the initial flow dominated distribution pattern for at least 60 min. The Iodine-123 isotope has been complexed with amines of the amphetamine type[8], while the Tc99 m isotope has been incorporated into a lipophilic molecule with d,1-hexamethylene-propyleneamine oxime (HMPAO)[14]. The pharmacokinetics of this compound have been analysed recently[11]. Tc99 m is much easier to handle in the laboratory as compared to Iodine-123 and Tc99 m is also cheaper. The physical halflife is 6.1 hours. It decays to Tc99 with the emission of a single gamma ray (140 KeV), which is optimal for tomographic imaging by SPECT. The isotope is available daily.

Clinical Study

Patients: At present our study population comprises 28 patients (age 29 ± 11, mean $\pm$ SD) with drug-resistant partial epilepsy characterized by frequent seizures, despite optimal antiepileptic treatment, including new drugs under investigation such as vigabatrin and oxcarbazepine. They exhibit simple or complex partial seizures with or without secondary generalization. The patients have been studied by SPECT as part of a presurgical evaluation programme. An interim report of this series has been published (Andersen *et al.*, 1988).

Ictal Studies: When the interictal neuro-imaging studies had been performed the patient data were analyzed and discussed. If the data were incongruent, ictal EEG and SPECT studies were performed. The attacks were provoked by reducing the anti-epileptic medication over a few days before the stipulated day of ictal SPECT and EEG-measurement during video observation. The Tc99 m-HMPAO was injected during the attack and the patient brought to the SPECT unit after intensive anti-convulsant therapy had been instituted, that is, within 1–1.5 hours.

Regional CBF Measurements: Regional CBF was measured by a brain dedicated SPECT-scanner, TOMOMATIC 64 (Medamatic Inc, Copenhagen) using xenon-133 inhalation and Tc99 m-HMPAO injection. First the xenon-133 inhalation study was performed. Tc99 m-HMPAO was injected i.v. 1 hour later. Usually the patients were positioned so that a series of nine parallel slices at a 1.3 cm distance were obtained, situated at OM + 1 cm, OM + 2.3 cm up to OM + 11.6 cm. Using xenon-133, mean rCBF and the standard error was calculated in any region of interest, based on the clearance and the bolus distribution principle. Using xenon-133 and HMPAO the images were also evaluated visually without knowledge of the case

history or other patient data and compared to the 200 other patient studies performed by our group during the same period. CT and MR scanning was performed using a Siemens CT scanner (Somatom DRG), and Magnetom[r] respectively. MRI was not performed in patients with a clearcut CT-lesion in the temporal lobes.

EEG-recordings: Sixteen-channel scalp recordings (Siemens-Elema) were obtained, using a standard 10/20 electrode placement including zygomatic or sphenoidal electrodes. Sleep-recordings and thiopental activation were performed in all patients. No intracranial recordings were performed.

Results

In 24 of the 28 patients studied so far, we have observed an obvious low flow region in one temporal lobe using Tc99 m-d, 1-HMPAO. In the remaining four patients the lesions were borderline and thus only tentative side-localization could be concluded from the SPECT study. In one of these patients EEG showed a corresponding localization. In another case, CT and MRI corresponded with SPECT while the EEG indicated that the contralateral temporal lobe contained the epileptogenic focus. An ictal SPECT and EEG study is planned in this case. A third patient had a monolateral congruent ictal and interictal study using SPECT but the EEG showed bilateral slow activity with no preponderance. CT and MRI were normal. In the last patient no focal hypoperfusion was seen using Tc99 m-HMPAO and the EEG was inconclusive with respect to side as well. CT and MRI showed a small lesion in the upper posterior part of the right parietal lobe (case 83).

The low flow regions in the 24 patients were well demarcated from normal tissue and involved in many cases higher (parietal and frontal) ipsilateral regions of the brain. In 13 patients a smaller region of low flow was observed in the contralateral (mirror) temporal lobe.

The perfusion asymmetry of the temporal lobes was obvious in 16 of the 27 patients studied by xenon-133 inhalation SPECT. In a further 6 patients a tentative localization of the side could be made from the xenon-133 images. The xenon-133 diagnosis was in agreement with the result obtained by HMPAO in 21 cases, while in the 22nd case (case 83) xenon-133 suggested the correct side of the focus.

SPECT was superior to CT as well as MRI for visualising the focal epileptogenic region suggested by the EEG. A focal abnormality was seen on 16 of the CT-scans. It was confirmed by the other neuroimaging tools in 15 patients, while one CT scan probably showed an irrelevant change. In another 11 patients no structural lesion was seen on the CT scan. MRI has so far been performed in 21 patients. Congruence with

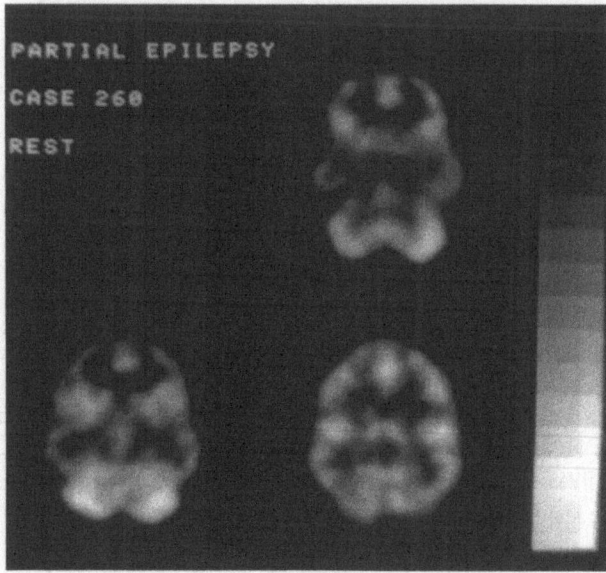

Fig. 1. Interictal SPECT images of rCBF. Case 260 studied by Tc99 m-HMPAO injection. Left is situated to the left and the nose of the patient is up. The blood flow distribution in the horizontal plane situated 2.3 cm, 3.6 cm, and 5 cm above the orbito meatal plane are given. A low flow region is evident in the left temporal lobe, but the side to side asymmetry is not very impressive

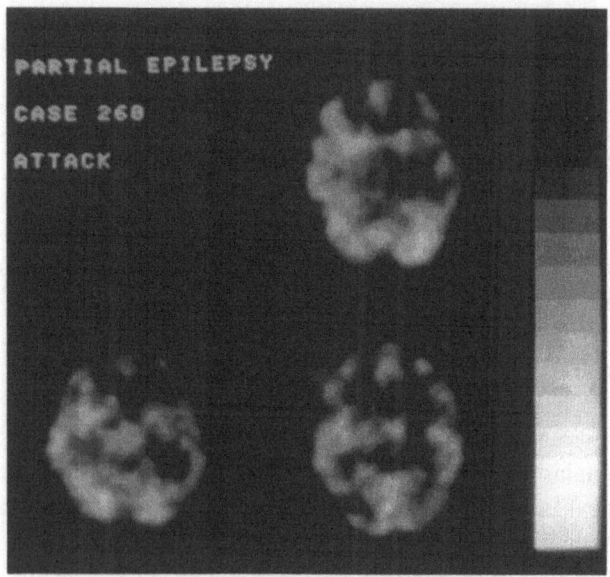

Fig. 2. Case 260 studied during a complex partial attack. The distribution of CBF at the orbito-meatal plane + 2.3 cm, + 3.6 cm and + 5 cm are given. The very high blood flow values in the left frontotemporal region are evident. They corresponded to a spike focus measured by EEG at the time of injection. MRI and CT − scan did not detect significant focal changes

the functional tools was found in 11 cases. MRI was inconclusive with respect to side localization in 8 patients and was probably false positive in 1. In 5 of the MRI and CT negative cases a consistent diagnosis was made by combining the results of the functional tools. So far, side localization has been impossible in one patient only.

In seven patients where structural and functional data were incongruent or inconclusive, the interictal studies were supplemented by ictal EEG and SPECT recordings. In 5 of these cases the operation was a success. The rest have not been operated yet. Figs. 1 and 2 shows a typical patient where the low flow region is evident interictally using Tc99 m-HMPAO. The defect is best seen in the high resolution Tc99 m-HMPAO study. The same patient was studied during an attack (Fig. 2) − CT and MRI were normal.

A total of 16 patients have been operated on to date. Details will be presented elsewhere. All patients had benefit from the intervention with only very few or no seizures after surgery. Surgery was performed through a fronto-temporal craniotomy to achieve a subfrontal approach; the medial part of the Sylvian fissure could then be opened and the uncus exposed. A small corticotomy was made and the uncus, amygdala and medial 1–1.5 cm of hippocampus were removed. One of the patients, case 109, in whom CT and MRI as well as EEG were inconclusive, was operated on the basis of the SPECT scans alone.

In 11 patients the SPECT result and the case history corresponded with respect to side localization. No false positive indication of side was given by any of the other 17 patients.

Discussion

The results correspond closely to those presented by Mazziotta and Engel[13] using PET. Our results are also in accordance with previous comparative investigations of EEG and SPECT in patients with partial epilepsy (Stefan *et al.*[20]). In our series the long-term success cannot be assessed yet, but the short term success is evident: all 16 operated patients have been relieved of most of their very frequent and otherwise intractable attacks.

Using the I-123 HIPDM and SPECT Lee *et al.* studied 16 patients with DRTLE[12]. Ictal HIPDM localized epileptic foci in 13 of 14 patients with a unilateral temporal focus and provided confirmative evidence of the epileptic focus in 11 patients by demonstrating maximally increased rCBF in epileptic foci that had shown decreased activity in a previous interictal study. The correlation with simultaneously recorded ictal EEG provided further clues for localizing the epileptic focus.

The overall results from SPECT studies of CBF in temporal lobe epilepsy show an average detection rate of significant focal rCBF changes in about 75% of the patients[2, 6, 12, 17, 18, 20]. The use in generalized epilepsy is controversial and still doubtful. The preliminary experience shows that, in many patients with DRTLE in whom there is doubt with regard to localization of the seizure-eliciting area of the brain, the site of the epileptogenic focus can be identified by SPECT. As a consequence, surgical therapy may be offered to more of these patients, if they fulfill the clinical criteria for this therapy. SPECT or PET scanning would seem to be indispensible for the optimal presurgical work-up of patients with partial epilepsy.

Acknowledgements

This study was supported by the Danish Medical Research Council and by the Danish Hospital Foundation for Medical Research and the Lundbeck foundation.

References

1. Andersen AR, Friberg L, Olsen TS, Olesen J (1988) SPECT demonstration of delayed hyperemia following hypoperfusion in classic migraine. Arch Neurol 45: 154–59
2. Andersen AR, Gram L, Kjær L *et al* (1988) SPECT in partial epilepsy. Identifying side of the focus. Acta Neurol Scand 45 [Suppl 117]: 90–94
3. Brodersen P, Paulson OB, Bolwig TG (1973) Cerebral hyperemia in electrically induced epileptic seizures. Arch Neurol 28: 334–338
4. Celsis P, Goldman T, Henriksen L, Lassen NA (1981) A method for calculating regional cerebral blood flow from emission computed tomography of inert gas concentrations. J Comput Assist Tomogr 5: 641–645
5. Engel J (1988) The role of neuroimaging in the surgical treatment of epilepsy. Acta Neurol Scand 45 [Suppl 117]: 84–89
6. Gjerstad L, Nyberg-Hansen R, Taubøll E *et al* (1987) Some aspects of regional cerebral blood flow (rCBF) in epilepsy, using single photon emission computer tomography (SPECT) (1987). In: Dam M, Johannessen SI, Nilsson B, Sillanpää M (eds) Epilepsy: Progress in treatment. John Wiley & Sons, Chichester, pp 69–80
7. Ingvar DH (1973) Regional cerebral blood flow in focal cortical epilepsy. Stroke 4: 359–360
8. Kuhl DE, Barrio JR, Huang SC *et al* (1982) Quantifying local cerebral blood flow by N-isopropyl-p (I-123) iodo amphetamine tomography. J Nucl Med 23: 196–203
9. Latack JT, Abou-Khalil BW, Siegel GJ, Sackellares JC, Ga-

brielsen TO, Aisen AM (1986) Patients with partial seizures: Evaluation by MR, CT and PET imaging. Neuroradiology 159: 159–163
10. Lassen NA, Sveinsdottir E, Kanno I, Stokely EM, Rommer P (1978) A fast moving single photon emission tomograph for regional cerebral blood flow studies in man. J Comput Assist Tomogr 2: 661–662
11. Lassen NA, Andersen AR, Friberg L, Paulson OB (1988) The retention of Tc99 m-d, 1-HMPAO in the human brain after intra-carotid bolus injection; a kinetic analysis. J Cereb Blood Flow Metab 8 [Suppl 1]: S 13–S 22
12. Lee BI, Markand ON, Wellman HM *et al* (1988) HIPDM-SPECT in patients with medically intractable complex partial seizures. Ictal study. Arch Neurol 45: 397–402
13. Mazziotta JC, Engel J (1984) The use and impact of positron computer tomography scanning in epilepsy. Epilepsia 25 [Suppl 2]: S 86–S 104
14. Neirinckx RD, Canning LR, Piper IM *et al* (1987) Technetium-99 m d, 1-HMPAO: A new radiopharmaceutical for SPECT imaging of regional cerebral blood perfusion. J Nucl Med 28: 191–202
15. Penfield W, Von Santha K, Cipriani A (1939) Cerebral blood flow during induced epileptiform seizures in animal and man. J Neurophysiol 2: 257–267
16. Plum F, Posner JB, Troy B (1938) Cerebral metabolic and circulatory responses to induced convulsions in animals. Arch Neurol 18: 1–13
17. Ryding E, Rosen I, Ingvar DH (1988) SPECT measurements with Tc99 m-HMPAO in focal epilepsy. J Cerebr Blood Flow Metab 8: S 95–S 100
18. Sanabria E, Chauvel P, Askienazy S *et al* (1983) Single photon emission computed tomography (SPECT) using I-123-isopropyliodo-amphetamine (IAMP) in partial epilepsy. In: Baldy-Moulinier *et al* (eds) Current problems in epilepsy: Cerebral blood flow, metabolism and epilepsy. John Libbey, London, pp 82–87
19. Smith FW, Gemmel HG, Sharp PF (1987) The use of Tc99 m-d, 1-HMPAO for the diagnosis of dementia. Nudl Med Commun 8: 525–33
20. Stefan H, Kuhnen C, Biersack HJ, Riechmann K (1987) Initial experience with 99 m Tc-hexamethyl-propylene amine oxime (HM-PAO) single photon emission computer tomography (SPECT) in patients with focal epilepsy. Epilepsy Res 1: 134–138
21. Theodore WH, Newmark ME, Sato S *et al* (1983) F-18-fluorodeoxyglucose positron emission tomography in refractory complex partial seizures. Ann Neurol 14: 429–437
22. Vorstrup S (Thesis) (1988) Tomographic cerebral blood flow measurements in patients with ischemic cerebrovascular disease and evaluation of the vasodilatory capacity by the acetazolamide test. Acta Neurol Scand [Suppl 114] 77: 1–48

Correspondence: A. R. Andersen, M.D., Department of Neurology, N2081 Rigshospitalet, Blegdamsvej 9, DK-2100 Copenhagen, Denmark.

Acta Neurochirurgica, Suppl. 50, 84–87 (1990)

Positron Emission Tomography Findings Relevant to Neurosurgery for Epilepsy

G. Pawlik, V. A. Holthoff, J. Kessler, J. Rudolf, I. R. Hebold, J. Löttgen, and **W.-D. Heiss**

Max-Planck-Institut für neurologische Forschung, Cologne University Hospital, Department of Neurology and Psychiatry, PET Laboratory, Köln, Federal Republic of Germany

Summary

Using the 2-[F-18]fluorodeoxyglucose method, 213 positron emission tomographic (PET) studies of local brain glucose metabolism (CMRglu) were performed in 124 patients with various forms of epilepsy. Interictal PET scans of primary epileptics typically showed some global metabolic depression and decreased functional activity of insular, basal and anterior temporal cortex. Epilepsia partialis continua Kozevnikov was characterized by hypo- or hyper-metabolism of perirolandic cortex. Tuberous sclerosis was distinguished by neocortical foci of significantly decreased glucose consumption. Even in the interictal resting state, with regard to sensitivity (> 90%) and accuracy of focus localization, PET was superior to other diagnostic methods in typical temporal lobe epilepsy. Averaging 23% below normal CMRglu, the majority of hypometabolic foci were found in mesial temporal structures. Improved distinction between the epileptogenic area and the surrounding tissue showing comparatively normal functional responsiveness, was achieved by psychophysical activation using emotional speech or continuous visual recognition during PET scanning. In patients who had undergone total cerebral hemispherectomy because of uncontrolled epilepsy, remarkable recruitment of association areas was observed on both motor and speech activation.

Keywords: Epilepsy; Positron Emission Tomography; cerebral glucose metabolism; functional activation.

Introduction

Epidemiological studies consistently demonstrate that a large proportion of all epilepsies are caused by some focal abnormalities of brain function at certain sites of predilection. According to recent estimates, partial seizures account for over 60% of all cases of epilepsy, and about 60% of these patients are refractory to drug therapy[8]. It is in these cases that neurosurgical treatment aimed at seizure control without producing additional neurological deficit may be considered. However, for this approach to be efficacious, distinct localization of the epileptogenic focus is essential, which can be attempted in several ways. Among the currently applied technical methods, positron emission tomography (PET) is distinguished by its unique capability of noninvasive, quantitative imaging in three dimensions and at reasonable spatial resolution of well-defined parameters of normal or impaired brain physiology, e.g., energy metabolism, blood flow, and receptor densities. This contribution summarizes our experience with the imaging of brain glucose metabolism (CMRglu), using the 2-[F-18]fluorodeoxyglucose method and PET[7] in various groups of patients presenting with recurrent seizures as their prominent clinical problem.

Patients and Methods

A total of 213 PET studies (7 ictal, 123 in interictal resting condition, 83 during psychophysical activation) were performed in 124 patients of whom 14 had idiopathic, 107 had symptomatic epilepsy, and 3 were seizure-free, hemispherectomized previous epileptics. In each case, the clinical diagnosis was supported by typical EEG changes. From all patients, x-ray computed tomography (CT) and/or magnetic resonance images (MRIs) were available for comparison, and a comprehensive neuropsychological test profile was obtained.

Using a previously described procedure[4, 12] and a 7-slice PET scanner (Scanditronix PC 384) with an in-plane resolution of 7.8 mm at a slice thickness of 11 mm, each PET study provided 14 original, partially overlapping, transaxial CMRglu images extending from the canthomeatal plane to the supraventricular level, permitting subsequent tomographic image reconstruction at any desired angle. Regional metabolic quantitation was achieved by means of a semiautomatic mapping routine[5], and a block shifting algorithm[6] was used to improve the anatomical comparability among studies.

Results

Idiopathic Epilepsy

In the interictal resting state, patients with idiopathic epilepsy typically showed low normal (grand mal) or

CT   MRi
T₁   T₂

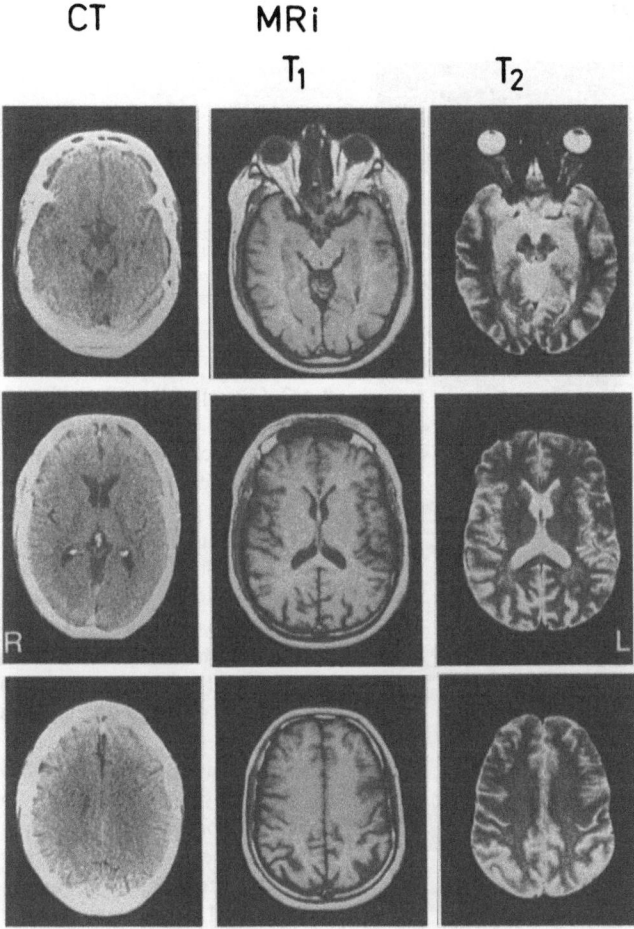

Fig. 1. CT and (approximately) corresponding T 1- and T 2-weighted MRI scans across inferior temporal lobes, thalamus, and oval center, respectively, of a 33-year-old male patient with drug-resistant simple, complex partial, and generalized tonic-clonic seizures since age 16, showing no significant morphologic abnormality

subnormal (petit mal) global CMRglu, depending also on their current anticonvulsant medication. Furthermore, there was some mild, widespread functional depression of insular, anterior and basal temporal cortex bilaterally that was insufficiently compensated during spontaneous speech. Only in grand mal epileptics was additional mild hypometabolism occasionally found in various neocortical and subcortical regions.

Symptomatic Epilepsy

Among the epilepsies of focal origin, the least consistent PET findings were obtained in 4 patients with *Kozevnikov epilepsy*, showing both some global metabolic depression and, unilaterally, either distinct focal hypermetabolism of precentral and supramarginal cortex, or regional metabolic decreases of perirolandic, basal frontolateral, anterior and mesial temporal cortex.

One or several neocortical foci of significant hypometabolism, mostly lacking any corresponding abnormality in CT, were consistently found[10] in 9 patients with *tuberous sclerosis* and tonic-clonic seizures. In 10 patients with typical *temporal lobe epilepsy*, a comparison of various modern neuroimaging methods (CT, MRI, SPECT, PET) and 16-channel EEG longterm recordings including sphenoidal leads and video monitoring revealed significant superiority[9] of resting PET with regard to the sensitivity and accuracy of localization of the interictally hypometabolic epileptogenic focus that did not necessarily coincide with the area of increased activity, as seen in ictal PET scans.

In the entire series of 87 patients with complex partial seizures studied in the interictal resting state, the focal depression of CMRglu averaged -23%, as compared to healthy control subjects matched for age and sex. Using single-observation t-tests, in 79 (approx. 91%) of those patients a focal abnormality of the metabolic pattern was detected. The vast majority of those foci were found in mesial temporal structures comprising temporal pole, uncus, amygdala, parahippocampal and hippocampal gyri. Fifty-two percent were unilateral, 48% bilateral, mostly (32%) showing some asymmetry of size and of the severity of metabolic depression. On the side of the (larger) focus, the amygdala could be distinguished in 46% of cases. However, because the resting CMRglu of mesial temporal structures is also low in non-epileptic subjects, the contrast between abnormal and relatively normal tissue often was quite poor, thus precluding a clear definition of the outlines of the focus for selective resection.

Therefore, *neuropsychological activation* paradigms were applied for the functional contrast enhancement between the presumably unresponsive epileptogenic lesion and the surrounding areas actively participating in task performance. In addition to their resting study, 61 patients (52 right-handers, 9 left-handers) underwent PET scanning during emotional spontaneous speech, and 15 of them had another PET study while performing a mixed verbal/nonverbal continuous visual recognition task, with the level of difficulty adjusted according to the individual's mnemonic abilities. On average, speech produced a 20% increase in global CMRglu, showing no effect of handedness or of side of focus. The regional metabolic activation, however, was significantly heterogeneous, with the smallest changes being observed in the epileptogenic focus (12%), and the largest (34%) in cerebellum, sensorimotor, premotor, Wernicke's, and Broca's areas. The metabolic contrast between the focus and reference

REST   SPEECH   CVR

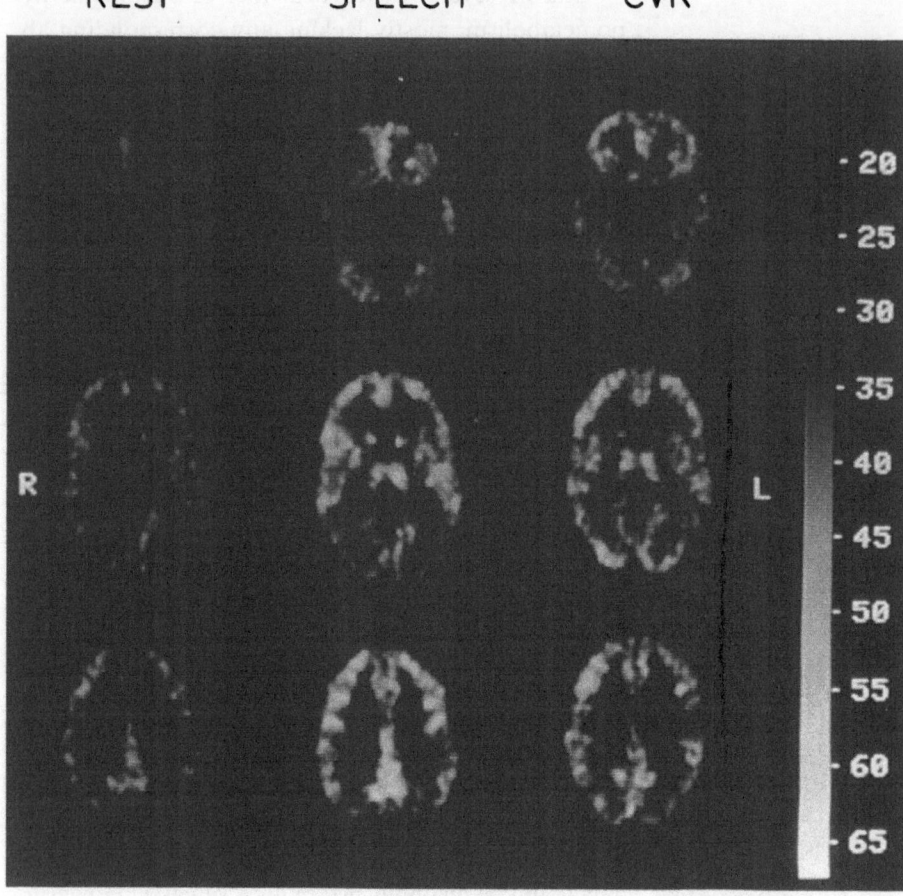

Fig. 2. Corresponding FDG-PET images of same patient as in Fig. 1, showing brain glucose metabolism (reference scale calibrated in µmol glucose consumption per minute per 100 g brain tissue) in standard resting condition as well as during emotional speech and continuous visual recognition (CVR), respectively. In addition to the global depression of resting CMRglu that is most prominent in infratentorial and inferior temporal structures, there is persistent focal hypometabolism of polar and mesial temporal gray matter bilaterally, of right posterior thalamus, right posterior insular cortex, left frontal operculum, and left supramarginal and anterior parietal cortex

regions, e.g., the ipsilateral middle temporal gyrus (MTG), was thus increased by 30%, concurrently improving the visibility of the amygdala in 25% of cases. From rest to speech, only the left-handers showed a strong functional shift to the right in Wernicke's area, while the right-handers exhibited significantly greater lateralization to the left hemisphere in premotor cortex and Broca's area, and to the right in cerebellum.

Continuous visual recognition produced a similar overall effect and only slightly poorer focus-to-MTG contrast, but the pattern of regional metabolic increases was significantly different, showing maximum activation in visual and frontolateral cortex. To date, selective focus resection has been performed in 10 patients evaluated in this way, and during the observation period of 7 to 27 months all have remained free of seizures.

Hemispherectomy

After 4 to 36 seizure-free postoperative years, three patients who had undergone total cerebral hemispherectomy (2 right, 1 left) at the ages of 9, 18, and 27, respectively, because of medically intractable, symptomatic seizures, were studied with PET both in a resting state and while performing coordinated, segmental limb movements on the paretic side, in order to investigate the brain's capacity for functional recovery following extensive neurosurgery. One patient who had been studied also preoperatively, exhibited a remarkable increase in global CMRglu. All patients' resting metabolism was in the lower range of normal, showing some regional deactivation of basal temporal and parietal cortical areas. During motor activation, major metabolic increases were observed in ipsilateral prefrontal, premotor, and parietal association cortex, in basal ganglia, and in both cerebellar hemispheres. Only the patient with left hemispherectomy showed the largest activation in those right-hemisphere regions corresponding to Broca's and Wernicke's areas. An identical metabolic pattern was obtained in the same patient's PET speech study.

Conclusion

The present findings clearly demonstrate that PET has a distinct place in the presurgical evaluation of focal epilepsies. Similar PET results on brain energy metabolism and blood flow in the interictal resting state have been reported from several centers[1,2,11,13]. Likewise, focal increases in opiate receptor density were observed in temporal lobe epilepsy[3]. Although some regional abnormality of interictal metabolism may also be found with primary epilepsies, these changes typically are less severe and spatially not as limited as in focal epilepsy. Furthermore, the focus detection characteristics of PET can be significantly improved by appropriate psychophysical activation, thus permitting greater selectivity and enhanced efficacy of neurosurgical treatment. As illustrated by the PET studies of hemispherectomized patients, various rather unexpected compensatory neuronal mechanisms may serve to minimize even the neurologic deficit resulting from late, extensive neurosurgery for epilepsy.

References

1. Bernardi S, Trimble MR, Frackowiak RSJ, Wise RJS, Jones T (1983) An interictal study of partial epilepsy using positron emission tomography and the oxygen-15 inhalation technique. J Neurol Neurosurg Psychiatry 46: 473–477

2. Engel J Jr. Kuhl DE, Phelps ME, Mazziotta JC (1982) Interictal cerebral glucose metabolism in partial epilepsy and its relation to EEG changes. Ann Neurol 12: 510–517

3. Frost JJ, Mayberg HS, Sadzot BL, Fisher RS, Dannals RF, Wilson AA, Lever J, Ravert HT, Wagner HN Jr (1988) Quantitation of changes in opiate receptors measured by C-11-diprenorphine and carfentanil in epilepsy patients by PET. J Nucl Med 29 [Suppl]: 796–797

4. Heiss W-D, Pawlik G, Herholz K, Wagner R, Göldner H, Wienhard K (1984) Regional kinetic constants and cerebral metabolic rate for glucose in normal human volunteers determined by dynamic positron emission tomography of [18F]-2-fluoro-2-deoxy-D-glucose. J Cereb Blood Flow Metab 4: 212–223

5. Herholz K, Pawlik G, Wienhard K, Heiss W-D (1985) Computer assisted mapping in quantitative analysis of cerebral positron emission tomograms. J Comput Assist Tomogr 9: 154–161

6. Pawlik G, Herholz K, Wienhard K, Beil C, Heiss W-D (1986) Some maximum likelihood methods useful for the regional analysis of dynamic PET data on brain glucose metabolism. In: Bacharach SL (ed) Information processing in medical imaging. Martinus Nijhoff, Dordrecht Boston Lancaster, pp 298–309

7. Reivich M, Kuhl D, Wolf A, Greenberg J, Phelps ME, Ido T, Casella V, Fowler J, Hoffman E, Alavi A, Som P, Sokoloff L (1979) The (18F)fluorodeoxyglucose method for the measurement of local cerebral glucose utilization in man. Circ Res 44: 127–137

8. Schomer DL (1983) Current concepts in neurology, partial epilepsy. N Engl J Med 309: 536–539

9. Stefan H, Pawlik G, Böcher-Schwarz HG, Biersack HJ, Burr W, Penin H, Heiss W-D (1987) Functional and morphological abnormalities in temporal lobe epilepsy: a comparison of interictal and ictal EEG, CT, MRI, SPECT and PET. J Neurol 234: 377–384

10. Szelies B, Herholz K, Heiss W-D, Rackl A, Pawlik G, Wagner R, Ilsen HW, Wienhard K (1983) Hypometabolic cortical lesions in tuberous sclerosis with epilepsy: demonstration by positron emission tomography. J Comput Assist Tomogr 7: 946–953

11. Theodore WH, Newmark ME, Sato S, De LaPaz R, DiChiro G, Brooks R, Patronas N, Kessler RM, Manning R, Margolin R, Channing M, Porter RJ (1984) 18F-fluorodeoxyglucose positron emission computed tomography in refractory complex partial seizures. Ann Neurol 14: 429–437

12. Wienhard K, Pawlik G, Herholz K, Wagner R, Heiss W-D (1985) Estimation of local cerebral glucose utilization by postron emission tomography of [18F]2-fluoro-2-deoxy-D-glucose: a critical appraisal of optimization procedures. J Cereb Blood Flow Metab 5: 115–125

13. Yamamoto YL, Ochs R, Gloor P, Ammann W, Meyer E, Evans AC, Cooke B, Sako K, Gotman J, Feindel WH, Diksic M, Thompson CJ, von Robitaille Y (1983) Patterns of rCBF and focal energy metabolic changes in relation to electroencephalographic abnormality in the inter-ictal phase of partial epilepsy. In: Baldy-Moulinier M, Ingvar D-H, Meldrum BS (eds) Current problems in epilepsy: cerebral blood flow, metabolism and epilepsy. John Libbey, London Paris, pp 51–62

Correspondence: G. Pawlik, M.D., Max-Planck-Institut für neurologische Forschung, Cologne University Hospital, Department of Neurology and Psychiatry, PET Laboratory, Joseph-Stelzmann-Strasse 9, Bldg. 30, D-5000 Köln 41, Federal Republic of Germany.

Acta Neurochirurgica, Suppl. 50, 88–94 (1990)
© by Springer-Verlag 1990

Quantitative Analysis of 18/FDG-PET in the Presurgical Evaluation of Patients Suffering from Refractory Partial Epilepsy

Comparison with CT, MRI, and Combined Subdural and Depth. EEG

R. M. Chr. Debets[2], C. W. M van Veelen[1], P. Maquet[3], A. C. van Huffelen[1], W. van Emde Boas[2], B. Sadzot[3], J. Overweg[2], D. N. Velis[2], D. Dive[3], and G. Franck[3]

[1] University Hospital Utrecht, Department of Neurosurgery and Clinical Neurophysiology, Utrecht, the Netherlands
[2] Instituut voor Epilepsiebestrijding, "Meer en Bosch-De Cruquiushoeve", Heemstede, The Netherlands
[3] University of Liège, Department of Neurology and Cyclotron Research Centre Liège, Belgium

Summary

CT, MRI, 18/FDG-PET and Depth. EEG, performed with subdural and depth electrodes were part of the presurgical evaluation in 22 patients. Statistical analysis of 18/FDG-PET was performed to compare cerebral utilization of glucose to that of normal age matched controls. The findings of CT, MRI, and quantitative analysis of PET are compared with those of ictal Depth. EEG. A positive correlation between CT and Depth. EEG was obtained in 23% of the patients and between MRI and Depth. EEG. in 50%. For both imaging techniques a negative correlation was found in 5%.

Regional abnormalities were found with quantified PET in 95% of the patients and were concordant with Depth. EEG. for side of onset in 77% of the patients and for lobe of onset in 59%. A possibly false localising PET result for lobe of onset was obtained in 8 patients (36%). Limitations of PET were most apparent in patients with regional mesiolimbic or bilateral seizure onset. A favourable outcome of surgery was associated usually with positive convergence of both methods.

PET may be a valuable contribution to the research and management of partial complex epilepsy, but at present cannot be considered a reliable alternative to invasive EEG methods in patients without clear unilateral focus localization on surface EEG.

Keywords: Epilepsy; CT; MRI; 18/FDG-PET; PET; Subdural and intracranial electrodes.

Introduction

In patients suffering from medically intractable partial complex seizures, resective surgical treatment can be considered only if a focal origin of the seizures can be demonstrated, located in a part of the brain that can be removed without causing unacceptable neurological deficit. Among the usual prerequisites of the initial presurgical evaluation in these patients are a careful analysis of the seizure history and a clinical neuropsychological assessment, CT and MRI imaging and extensive interictal and ictal EEG studies, including closed circuit-T.V. seizure monitoring and the application of special semi-invasive electrodes. In patients in whom no specific causal structural lesion, expanding or otherwise, is demonstrated, the further course of action in principle is determined by the degree of convergence of the available data. However controversy exists about the number of presurgical investigations that are needed to warrant surgery and about the reliability of the individual investigations for prediction of outcome. The most reliable method appears as yet to be the intracranial recording of seizures. As this method is invasive clearly the need still remains for reliable non-invasive methods.

The determination of cerebral glucose metabolism with Positron Emission Tomography (PET) with [18F] 2 fluoro-2-deoxy-D-glucose (18/FDG) has been shown to be a valuable addition to the presurgical assessment of patients with partial complex seizures[2, 3, 4]. 22 patients, already selected for depth EEG monitoring (D.EEG) as part of their presurgical screening, were studied with 18/FDG-PET in addition to prior imaging studies (CT and MRI). The decision concerning eventual surgical treatment in these 22 patients depended essentially on the ictal D.EEG findings. The present investigation was designed to establish the potential value of standardized statistical regional analysis of 18/FDG metabolism in PET as an adjunct or even alternative to Depth. EEG. The convergence of the various studies in all 22 patients and their value for the prediction of surgical outcome is evaluated.

Patients and Methods

Patients

Between 1985 and 1988 22 patients with medically intractable complex partial seizures were identified, in whom ictal EEG and closed circuit T.V. monitoring failed to establish an unequivocal localization or lateralization of their seizure onset and who accordingly were selected for Depth EEG monitoring as a final step in their presurgical work-up. There were 11 males and 11 females, aged 17–50 years (mean age 31 years). Of these 22 patients, 18 were considered eventually to be suitable candidates for surgery. Postoperative follow-up of at least six months was available for ten patients at the time of writing.

CT

Computerized Tomography was carried out, without and with contrast enhancement, with a Siemens Tomograph 310 or 350. Scans were obtained in transverse and coronal planes. Slice thickness and interleaf gap varied from 3–5 mm.

MRI

Magnetic Resonance Imaging (MRI) was performed with a super-conducting system (Teslacon R) operating at a field strength of 0.6 tesla (T). Transverse and sagittal scans were performed in all patients, while images in the coronal plane were included in selected cases. Slice thickness was 10 mm, with a 3 mm interleaf gap. In all cases T 1, T 2 and inversion recovery images were performed. Spin-echo images were collected with TR 500 msec and TE 32 msec (T 1 images) or alternatively with TR 3000 msec and a TE series with 32, 64, 96, 128 msec (T 2 images). The inversion recovery technique was obtained with TR 2100 msec, TE 32 msec and T1 500 msec.

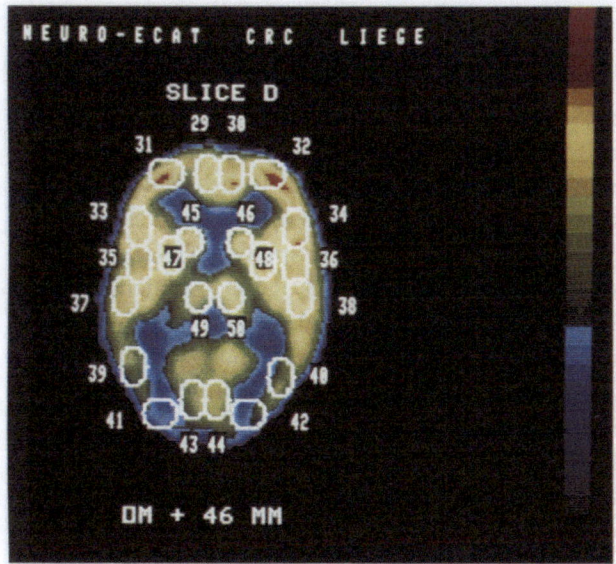

Fig. 1. Topographic representations of a number of regions of interest

18/FDG-PET

Positron Emission Tomography (PET) was carried out in patients during the interictal state, using the [18F]-fluoro-D-deoxyglucose method. A bolus of 5–10 mCi 18/FDG was injected intravenously. Arterial samples were drawn during the 45 minutes of the uptake period. The patient was in supine position on the table of the tomograph with the eyes closed and ears free with environmental light and noise reduced as much as possible. Six to 15 planes 8 mm apart were acquired parallel to the orbito-meatal plane (OM + 15 to OM + 80 mm) by a Neuro-Ecat (EG & ORTEC), shadow shields in, septa shields out, transverse FWHM:12.5 mm, axial FWHM:15 mm. Attenuation correction was calculated using an operator-positioned skull fitting ellipse algorithm and a uniform absorption coefficient of $0.088 \, cm^{-1}$. Cerebral metabolic rates were calculated using the operational equation of Sokoloff *et al.* (1977) modified by Phelps *et al.* (1979)[10], using the values of rate constants and "lumped constant" proposed by these authors for adult men. Regional values were obtained from 89, 29–35 pixel sized preselected regions of interest (2.23–2.69 cm²). They are illustrated in Figs. 1 and 2. Left right differences were calculated for all pairs of symmetrical regions following the formula $(L\text{-}R/L+R) \times 200$.

Metabolic indices were calculated by normalizing each regional metabolic rate by a mean cortical value considered as an intrinsic reference for each scan. This mean cortical value was calculated from 16 cortical regions appearing on the plane cutting through thalamic nuclei and basal ganglia. Metabolic indices thus provided a measure of regional variations of metabolic rates of glucose.

A control population of 12 subjects (mean age 26.3 ± 5.5) was studied in the conditions described above and used to define a confidence interval ($p < 0.05$) of interhemispheric differences and metabolic ratios.

The analysis of PET data was performed in a double blind way in two stages. First visual assessment disclosed one or several areas of decreased glucose utilization.

Subsequently in a quantitative comparative analysis with the values obtained in the normal population the statistical significance of the regional asymmetry of hypometabolism was assessed. No area of abnormal glucose utilization was considered significant unless it appeared in at least two separate planes and their left-right differences and/or metabolic indices differed significantly from normal values[5, 12].

D.EEG

Intracranial Depth EEG (D.EEG) monitoring was performed with a combination of subdural and intracerebral electrodes[6, 15]. Miniaturized flexible subdural multicontact electrodes, combined into bundles or reeds were introduced through two symmetrical fronto-central trephine holes (diameter 2.5 cm) and manipulated under fluoroscopy to cover most of the cerebral convexity. The number of individual electrodes varied from 5–12 per patient (mean = 9). In addition and using the same trephine-holes 2–4 depth electrodes were stereotactically implanted in the mesial temporal and/or mesial frontal structures. Each electrode has 7 recording leads. An extensive recording area and a detailed spatial delineation of the epileptogenic zone can thus be obtained from some 100 recording contacts with minimal risk to the brain. Anti-epileptic medication was usually tapered off during the EEG monitoring which continued for 5–12 days until a sufficient number of the patients' habitual seizures had been recorded. The onset of the EEG seizures were considered focal when ictal changes were initially confined to 1–2 adjacent contacts and

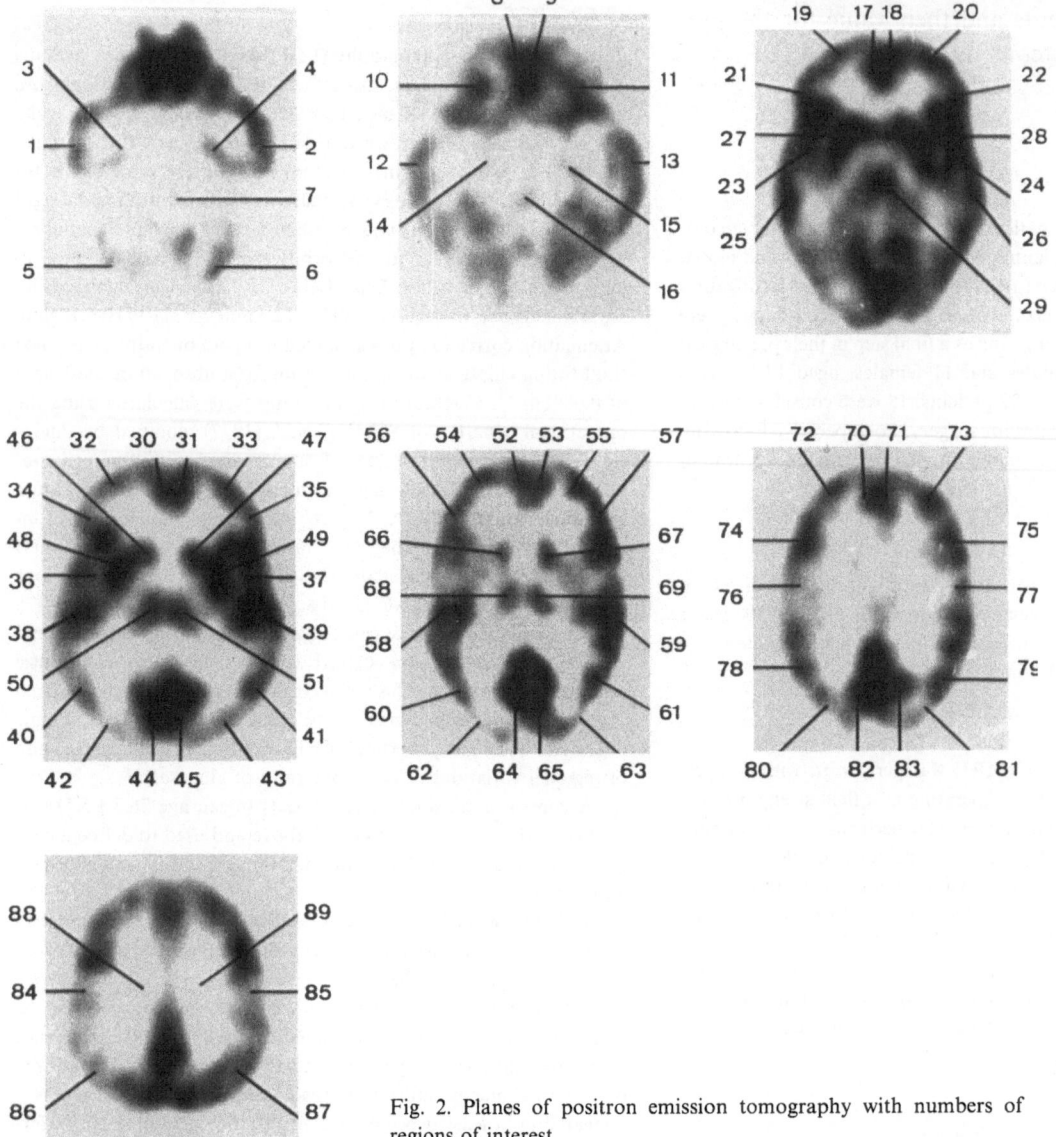

Fig. 2. Planes of positron emission tomography with numbers of regions of interest

regional when initially limited to 3–4 contacts on the same electrode or to the equivalent 1–3 most distal contacts on ipsilateral mesiolimbic electrodes. Initial changes simultaneously involving more than 4 adjacent contacts on one electrode or more than one subdural or two ipsilateral mesiolimbic electrodes were classified as widespread.

Results

The ictal D.EEG findings were classified according to the regions of surgical interest: a/focal frontal, b/regional mesio temporal, c/wide spread temporal including widespread neocortex and widespread mesio temporal + neocortex and d/bilateral mesio temporal.

The quantitative data obtained from the 18/FDG PET were related to the areas of metabolic interest which are indicated in Table 1. The CT and MRI data were correlated to these by visual assessment.

Table 2 illustrates the correlation of the onset of ictal D.EEG findings, and the areas of statistically significant metabolic abnormality of PET. The relative maximum of hypometabolism is also indicated. In 21 of the 22 patients investigated, decreased captation of 18/FDG was found. Accordingly correlation of side was positive in 17 and negative in 4 patients. In one patient the PET was normal. Negative correlation related to 3 patients with bilateral independent temporal foci without left or right preponderance. In one patient with a unilateral mesiotemporal onset of seizure activity, bilateral hypometabolism was found. The hypometabolism was maximal in the contralateral temporal lobe. In most patients who had seizure onset in the temporal lobe hypometabolism was found to be spreading over neighboring areas. The hypometabolism was

Table 1. *Regions of Interest as Lumped in Lobar Areas from Seven Planes of Reference and 89 Preselected Regions of Interest Used in this Study*

Cortical regions	Medial frontal
	Lateral frontal
	Orbital frontal
	Inferior frontal
	Anterior frontal
	Middle temporal
	Superior temporal
	Parietal
	Lateral occipital
	Medial occipital
	Insula
	Sensorimotor cortex
	Hippocampus
Subcortical regions	Anterior striatum
	Caudate nuclei
	Lenticular nuclei
	Thalami
	Brainstem
	Cerebellum

considered primarily correlated with one lobe if the maximum area of hypometabolism was localised in that lobe. Correlation between D.EEG and PET concerning lobe of onset was positive in 13 patients and negative in 8. It should be noted that in only a few cases was hypometabolism found in the mesiotemporal region and that in 4 of the 8 patients who had mesiotemporal seizure onset in the D.EEG the maximum area of hypometabolism was outside the lobe of seizure onset in contrast to the 8 patients who had widespread temporal onset. In these patients only one showed maxium hypometabolism outside the temporal lobe displaying the onset of seizure activity.

As shown in Table 3, CT abnormalities were found in 6 patients. The localization was in agreement with the D.EEG finding in 5 of them for side and lobe. In one patient who had bilateral mesiotemporal onset in the D.EEG the CT was bilaterally abnormal in the frontal lobes. Correlation for hemisphere was positive in 6 and for lobe positive in 5 patients. The MRI of 10 patients was abnormal. The findings were in agreement with the results of D.EEG in 9 patients. In one patient with bilateral seizure involvement, correlation was negative.

A comparison was made between the positive findings of CT, MRI and 18/FDG-PET and their correlation as to the identification of side of lobe of onset

Table 2. *Correlation Between Seizure Onset in D. EEG and Regions of Hypometabolism of 18/FDG-PET Indicated by Statistical Analysis*

D. EEG	Frontal focal	Mesiotemporal focal + regional	Temporal widespread	Bilateral (mesiotemporal)
FDG-PET				
Sup. front.	o			
Lat. front.	× ×	o o ×	o o	×
Inf. front.		o o o ×	o o o o o	o
Orb. front.		o o × o	o o o o	o
Mesio temp.		o o	o	o
Ant/Lat. Temp.		× × □ o	o × × × ×	o
Mid. temp.		o o o □ o o o	o × × o o	o o
Sup. temp.		× × o o	o × o o o o o	o × ×
Insula		o o o o o	o o	
Central	o	o		
Parietal	×			
Occipital		o		

PET																							
Concordance side	+	+	+	+	+	+	+	−	+	+	+	+	+	+	+	+	+	+	−	−	−	+	+ tot = 17 / − tot = 4
Concordance lobe	+	+	−	+	+	+	+	−	−	−	+	+	+	+	+	+	+	+	−	−	−	−	+ tot = 13 / − tot = 8
Pat. no.	1	2	3	4	5	6	7	8	9	10	11	12	13	14	15	16	17	18	19	20	21	22	tot = 22

o = hypometabolism.
× = maximum of hypometabolism.
□ = contralateral hypometabolism.

Table 3. *Correlation Between Seizure Onset in D. EEG and Regions of Hypometabolism*

D. EEG	Frontal focal	Mesiotemporal focal + regional	Temporal widespread	Bilateral (mesiotemporal)	
CT/MRI					
Sup. front.	×			○ B	
Lat. front.	×			○	
Inf. front.					
Orb. front.					
Mesio temp.		○ × ×	× ○		
Ant/lat. temp.		○ × ×	○ ○ ○ ○	× B	
Mid. temp.					
Sup. temp.				×	
Insula					
Central					
Parietal				×	
Occipital					
CT					
Concordance side		+	+ + + +	+	+ tot = 6
					− tot = 1
Concordance lobe		+	+ + + +	−	+ tot = 5
					− tot = 1
MRI					
Concordance side	+ + +	+ +	+ + + + +	− +	+ tot = 11
					− tot = 1
Concordance lobe	+ + +	+ +	+ + + + +	+	+ tot = 11
					− tot = 1
Pat. no.	1 2 3	4 5 6 7 8 9 10	11 12 13 14 15 16 17 18	19 20 21 22	tot = 22

○ = abnormal CT.
× = abnormal MRI.
B = bilateral.

Table 4. *Comparison Between Positive Findings of CT, MRI, 18/FDG-PET and Depth EEG*

	CT	MRI	18 FDG-PET
Patients investigated			
n (%)	22 (100)	22 (100)	22 (100)
Abnormality found	6 (27)	12 (55)	21 (95)
Abnormality – DEEG side	6 (27)	11 (50)	17 (77)
Abnormality – DEEG Lobe	5 (23)	11 (50)	13 (59)

in the D.EEG. The results are presented in Table 4. CT appears to be the least sensitive with only 6 abnormal scans (27%) against 12 abnormal MRI investigations (55%) and 21 abnormal PET investigations (95%). Yet its specificity for side and lobe is almost identical to that of MRI (resp. 6 and 5 out of 6 respectively for side and lobe against 11 out of 12 positve findings) and higher than 18/FDG-PET with respectively 17 and 13 out of 21 investigations concordant for side and lobe. Overall the CT findings were positive

and in agreement with D.EEG both for lobe and side in 5 (23%) of the patients and suggestive of a wrong focus localization in one (5%). MRI results were in agreement with D.EEG in 11 patients and misleading in one (5%). Results of 18/FDG-PET were concordant with D.EEG in 59% of all patients for lobe of onset and in 77% for side. In only one patient was no abnormality found (5%). The findings were not in agreement with D.EEG either for side or for lobe of onset in 18% and 36% of all patients respectively.

Of the 22 patients investigated in the present series 18 turned out to be suitable candidates for surgery. Eleven patients had a classical temporal lobectomy, one a selective amygdala-hippocampectomy, one patient had a subtotal frontal lobectomy and in one a restricted frontal corticectomy were performed. Four patients are still awaiting surgery and four patients with bitemporal seizure onset have been rejected for surgery.

Within this selected population of 18 patients with unilateral seizure onset a positive correlation between

18/FDG-PET and D.EEG for lobe was present in 13 patients (72%) and for side even in 17 (94%). A wrong lateralization was suggested by PET in only one surgical candidate. However it should be noted that the retrospective selection bias was made possible by D.EEG and not by 18/FDG-PET itself. CT and MRI correctly localized the area of seizure onset in 5 (27%) and 10 (55%) respectively of these 18 patients without any falsely localising findings within this group of definite surgical candidates.

At the time of writing a post surgical follow-up period of more than 6 months (6 months − 4 years) is available for ten patients. Seven of them have remained seizure free, 2 are improved (> 75% seizure reduction) and one is unchanged. In 2 of these 10 patients PET findings did not agree with D.EEG. In one patient (pat.no 8) 18/FDG-PET showed bitemporal abnormalities and maximal hypometabolism contralateral to D.EEG onset and MRI abnormalities; following temporal lobectomy on the side indicated by D.EEG this patient has become seizure free. The other patient (pat.no 3) did not improve after temporal lobectomy indicated by D.EEG, CT, and MRI but not by 18/FDG-PET, which showed hypometabolism in the ipsilateral parietal (maximal) and central regions only.

Discussion

The epileptic focus in patients with partial epilepsy is characterized by a neuronal cell dysfunction, accompanied by morphological, physiological and biochemical changes. For the ultimate surgical treatment of these patients the exact identification of this focus is of paramount importance. In current practice, accurate localization essentially still belongs to the realm of EEG recording, the ictal depth EEG being more sensitive and therefore still indispensable in those patients in whom ictal surface recordings do not disclose unequivocal unilateral focal seizure onset. D.EEG by whatever method however remains an invasive procedure and a further search for non-invasive techniques for better localization of the area of possible surgical resection thus remains warranted.

In the present series, retrospectively the CT correctly identified the epileptogenic lobe in 22% and was misleading in only 5%. For MRI these numbers were 50% and 5% respectively. These findings appear comparable to those of others[13]. Although lesions demonstrated in CT or MRI are not necessarily identical with the epileptogenic site[9], both CT and MRI correlated very well with the lobe of seizure onset in all patients

with unilateral seizure onset in whom these abnormalities were found.

The potential value of 18/FDG − PET for the localization of partial epileptic foci has been demonstrated by Engel and others[2–5, 7, 9, 12, 13] and more precise results of this method may be obtained with the application of quantitative rather than visual analysis of the data[8, 14]. This approach may illustrate better than visual analysis the present limitations of 18/FDG-PET.

A. In PET a control population is used to define relevant interhemispheric differences and metabolic ratios. Due to the relatively high variance of data it is difficult to recognize asymmetry in hypometabolism when this is more or less equal in bilateral symmetric regions[7]. In the present study bilateral mesiotemporal involvement as found with D.EEG in 3 patients was not in agreement with the PET of these patients, which showed only unilateral involvement. Moreover in the one case that PET showed bilateral hypometabolism in symmetrical regions the epileptic focus was apparently localised in one temporal lobe as is illustrated by successful surgical treatment. Furthermore in this patient the maximum area of hypometabolism was pointing falsely towards the apparently non-affected side.

Proper lateralization of seizure onset in patients suspected of suffering from epilepsy with bilateral involvement appears to be difficult and may be obtained better with other methods[4, 16].

B. In the present study in only 3 out 16 patients (19%) with unilateral temporal lobe epilepsy was hypometabolism found in the mesiotemporal regions and in only 2 out of 8 patients (25%) with epilepsy of mesiotemporal onset. Maxima of hypometabolism were never found in that region. Given the spatial resolution of the present equipment it is generally not possible to measure the metabolic rate of glucose in the mesiotemporal region. At present 18/FDG-PET appears to have a restricted value in the evaluation of epilepsy of mesiotemporal onset and so far it has no place in selection of patients who are considered candidates for selective amygdala-hippocampectomy[17]. The finding that the maxima of hypometabolism were more often extralobar in mesiotemporal seizure onset than in widespread temporal onset remains unexplained and requires further evaluation.

C. When hypometabolism in 18/FDG-PET is multilobar it may be difficult to use PET to distinguish between complex partial seizures of temporal lobe or extratemporal origin[7]. This is clearly illustrated in the present study. As in this study quantified data were available, hypometabolism was considered related primarily to one lobe if the maximum area of hypometabolism was localised in that lobe. Quantification of hypometabolism thus performed may possibly increase the precision of the method and appear to be a valuable method of localising the lobe of seizure onset. It may provide a relevant additional parameter for choosing targets in D.EEG when indicated, particularly if the findings can be incorporated in a combined stereotactic approach[9]. However further evaluation of such a method of quantification is needed. From the present study it appears that CT and MRI when abnormal are better independent localizing methods when compared to D.EEG than PET.

18/FDG-PET cannot currently localize independently. If the method had been used as such in the present series, 8 of the 22 patients might possible have had surgery of the wrong brain region.

Recognizing the limitations of PET in bilateral seizure involvement and in cases of bilateral hypometabolism in symmetrical regions, the postoperative evaluation of 10 patients with a good outcome after 6 months − 4 years of follow-up revealed positive concordance between 18/FDG-PET and D.EEG in 9 of them and negative concordance correlated with lack of improvement in one patient. Quantitative 18/FDG-PET is likely to offer a major contribution to the clinical and scientific evaluation of patients with partial complex seizure disorders. At present it does not offer a reliable alternative to invasive EEG methods in patients without clear unilateral focus localization in surface EEG.

References

1. Abou-Khalil BW, Siegel GJ, Sackellare JC, Gilman S, Hichwa R, Marshall R (1987) Positron emission tomography. Studies of cerebral glucose metabolism in chronic partial epilepsy. Ann Neurol 22: 480–485
2. Engel J Jr, Kuhl DE, Phelps ME, Mazziotta JC (1982) Interictal cerebral glucose metabolism in partial epilepsy and its relation to E.E.G. changes. Ann Neurol 12: 510–517
3. Engel J Jr, Kuhl DE, Phelps ME, Crandall PH (1982) Comparative localization of epileptogenic foci in partial epilepsy by PCT and EEG. Ann Neurol 12: 529–537
4. Engel J Jr (1988) The role of neuroimaging in the surgical treatment of epilepsy. Acta Neurol Scand 78: 84–89
5. Franck G, Sadzot B, Salmon E, Depresseux JC, Grisar T, Peters JM, Guillaume M, Quaglia L, Delfiore G, Lamotte D (1986) Regional cerebral blood flow and metabolic rates in human focal epilepsy and status epilepticus. Adv Neurol 44: 935–948
6. Hulten K, van Lopez da Silva FH, Lommers JC, Storm van Leeuwen W, Van Veelen CWM, Vliegenthart WE (1977) Localization of epileptogenic areas in patients investigated with chronically indwelling electrodes using computer analyzes. Electroencephalogr Clin Neurophysiol 43: 533–567
7. Holmes MD, Kelly MA, Theodore WH (1988) Complex partial seizures, correlation of clinical and metabolic features. Arch Neurol 45: 1191–1193
8. Lancet, Editorial. SPECT and PET in Epilepsy, January 1989, pp 135–137
9. Olivier A, Marchand E, Peters T, Typer J (1987) Depth electrode implantation at the Montreal Neurological Institute and Hospital. In: Engel J (ed) Surgical treatment of the epilepsies. Raven Press, New York, pp 595–601
10. Phelps ME, Huang SC, Hoffman EJ, Selin C, Sokoloff L, Kuhl DE (1979) Tomographic measurement of local cerebral glucose. Metabolic rate in humans with (F-18) 2-fluoro-2-deoxy-d-glucose: validation of method. Ann Neurol 6: 371–388
11. Sperling MR, Sutherling WW, Nuwer MR (1986) New techniques for evaluating patients for epilepsy surgery. In: Engel J Jr (ed) Surgical treatment of the epilepsies. Raven Press, New York
12. Sadzot B, Salmon E, Maquet P, Debets R, Grisar Th, Lemarie Ch, Degueldre Ch, Franck G (1987) Assessment of interest of PET and MRI in complex partial epilepsy. J Cereb Blood Flow Metab 7 [Suppl 1]: S 432
13. Theodore WH, Dorwart R, Holmes M, Porter J, DiChiro G (1986) Neuro-imaging in refractory partial seizures: Comparison of PET, CT, and MRI. Neurology 36: 750–759
14. Theodore WH (1989) SPECT and PET in epilepsy. Lancet, march, pp 502–503
15. Veelen CWM, van Debets RMCh, Binnie CD (1986) Subdural and depth electrodes in preoperative screening for intractable seizures. Presented at the International Conference on Surgical Treatment of Epilepsy. Palm Desert, California, Febr. 20
16. Wieser HG, Elger CE, Stodeck SRG (1985) The "Foramen Ovale Electrode", a new recording method for the preoperative evaluation of patients suffering from mesio-basal temporal lobe epilepsy. Electroencephalogr Clin Neurophysiol 61: 314–322
17. Wieser HG (1986) Selective amygdala hippocampectomy: Indications, investigative technique and results. In: Symon L *et al* (ed) Advances and technical standards in neurosurgery, vol 13. Springer, Wien New York

Correspondence: C. W. M. van Veelen, M.D., Ph.D., University Hospital Utrecht, Department of Neurosurgery, Catharijnesingel 101, 3511 GV Utrecht, The Netherlands.

Acta Neurochirurgica, Suppl. 50, 95–99 (1990)
© by Springer-Verlag 1990

Identification of the Side of the Epileptic Focus with ^{123}I-Iomazenil SPECT

A Comparison with [18]FDG-PET and Ictal EEG Findings in Patients with Medically Intractable Complex Partial Seizures

A. C. van Huffelen[1], J. W. van Isselt[2], C. W. M. van Veelen[3], P. P. van Rijk[2], A. M. E. van Bentum[2], D. Dive[4], P. Maquet[4], G. Franck[4], D. N. Velis[5], W. van Emde Boas[5], and R. M. Chr. Debets[5]

[1] Department of Clinical Neurophysiology, University Hospital Utrecht, The Netherlands,
[2] Department of Nuclear Medicine, University Hospital Utrecht, The Netherlands,
[3] Department of Neurosurgery, University Hospital Utrecht, The Netherlands,
[4] Department of Neurology and Cyclotron Research Centre, University of Liege, Belgium
[5] Instituut voor Epilepsiebestrijding "Meer en Bosch", Heemstede, The Netherlands

Summary

[123]I-Iomazenil SPECT was performed in 17 patients who were considered candidates for surgery of epilepsy because of medically intractable complex partial seizures. In addition to this examination their presurgical evaluation consisted of long term ictal EEG-CCTV monitoring, CT, MRI and [18]FDG PET. In eight patients intracranial ictal EEG recordings were performed.

SPECT was assessed visually while PET data were analyzed quantitatively. Both SPECT and PET were compared to ictal EEG data and showed asymmetries in over 80% of patients in agreement with EEG findings. These three methods were in agreement in 65% of patients. SPECT showed abnormality contralateral to the EEG focus in one patient (6%) while PET always demonstrated ipsilateral dysfunction.

It is concluded that ^{123}I-Iomazenil SPECT may be considered a more economical and more widely available alternative to [18]FDG PET in the presurgical evaluation of patients with medically intractable complex partial seizures. In this respect ^{123}I-Iomazenil specifically reflects functional changes in the membranes of neurons while [18]FDG is related to glucose metabolism not only of neurons but also of glial cells.

Keywords: Epilepsy; epileptic focus; identification; SPECT; ^{123}I-Iomazenil; [18]FDG-PET; ictal EEG.

Introduction

Therapy through surgical removal of the epileptic focus should be considered in patients with medically intractable complex partial seizures. A prerequisite for such a treatment is the precise identification and delineation of the causative epileptic focus. At present ictal EEG recording with intracranial electrodes is considered the most reliable method of achieving this goal. Various less invasive techniques have been pursued to provide a suitable alternative.

With ^{18}F-fluoro-deoxy-D-glucose positron emission tomography ([18]FDG-PET) areas of decreased cerebral metabolic rate of glucose (CMR Glu) consistent with the lateralization or even localization of the primary epileptic focus have been demonstrated[3, 4, 9].

In brain areas from which complex partial seizures originated decreased cerebral blood flow has been shown with Hexa-methyl-propylene-amine-oxime single photon emission computed tomography (HMPAO-SPECT)[1, 6]. With the selective benzodiazepine (BZ) antagonist, Flumazenil, labelled with ^{11}C, a local reduction of BZ receptor binding has been shown in patients with complex partial seizures[7]. These BZ receptors form an integral part of the gamma-amino butyric acid (GABA)-chloride ionophore complex on the neuronal membrane. GABA opens the chloride channel and thus prevents cellular depolarisation. Since it is suggested that GABA-ergic dysfunction plays a major role in the epilepsies[2] this type of study may provide a more direct approach to the basic pathophysiology of the epileptic focus than studies which are concerned with glucose metabolism or with cerebral blood flow.

In the present study in patients with medically intractable complex partial seizures the demonstration of reduced BZ binding was evaluated in comparison with reduced glucose metabolism as shown with ^{18}FDG-PET and ictal EEG monitoring with depth and/or scalp electrodes. Reduced BZ binding was not investigated with the ^{11}C flumazenil PET technique, but with the less expensive and more widely available SPECT technique using ^{123}I-Iomazenil (a ^{123}I-flumazenil derivative) as a radioligand for BZ receptors.

Subjects and Methods

17 patients (14 men, 3 women) were investigated as part of their presurgical evaluation for medically intractable complex partial seizures. Their mean age was 34, SD 8 years. In all patients neuroimaging with CT and MRI was performed. No patients with a space occupying lesion were included in the study. In addition long term ictal EEG – closed circuit television (CCTV) monitoring, PET and SPECT were performed. Both PET and SPECT results were compared to the ictal EEG findings because ictal EEG was considered to provide the most reliable lateralization of the epileptic focus especially when using intracranial electrodes.

EEG

All patients were studied with long term interictal as well as ictal EEG and CCTV monitoring with conventional scalp electrodes as well as nasopharyngeal and/or sphenoidal electrodes. In addition, eight patients (47%) had prolonged intracranial EEG recordings with both subdural and depth electrodes. For this purpose generally 8–10 miniaturized flexible subdural reeds with 7 recording sites each were introduced through bilateral frontocentral trephine holes to cover frontal and centrotemporal parts of the cerebral convexity. In addition and using the same trephine holes 2–4 multi contact depth electrodes were stereotactically implanted bilaterally in hippocampus and/or amygdala.

PET

Interictal positron emission tomography (PET) was performed in all patients with a Neuro Ecat (in plane resolution FWHM 12 mm, transaxial resolution FWHM 15 mm) using the ^{18}F-fluoro-deoxy-D-glucose (FDG) technique. Six to 15 planes 8 mm apart and parallel to the orbito-meatal plane were measured. Cerebral metabolic rate for glucose (CMR Glu) was calculated according to the Sokoloff model modified by Phelps (1979). Areas of decreased glucose metabolism were analyzed quantitatively. Regional values were obtained for 89 regions of interest (2.25–2.69 cm^2) and calculated relative to the mean of cortical CMR Glu. Left-right differences relative to the mean were calculated according to the formula $2(L-R)/(L+R) \times 100\%$.

A control population of 12 subjects (mean age 26 years, SD 5.5 years) was studied in the same way. In order to be rated as hypometabolic an abnormal area had to be present in at least two adjacent separate planes. An area was considered as abnormal when it was significantly different from the normative data and/or showed a significant left-right difference (level of significance $p < 0.05$)

SPECT

Single Photon Emission Computed Tomography (SPECT) was performed interictally with the selective BZ antagonist ^{123}I-Iomazenil, derived from Flumazenil. In two patients on regular BZ medication this drug was withdrawn two weeks prior to the SPECT investigation.

Thirty minutes after i.v. injection of 400 MBq ^{123}I-Iomazenil, SPECT acquisition was started using a rotating gamma camera with a low energy parallel hole collimator (full angle rotation, 60 steps , 40 sec/view, in plane resolution FWHM 12 mm). SPECT images were reconstructed using a filtered back projection method. Twelve planes 9 mm apart and parallel to the orbito-meatal plane were measured. Attention was focussed on the three lower planes providing data on frontobasal and temporal regions. SPECT images were visually assessed for major asymmetries. No such asymmetries were found in a control population consisting of 9 subjects (mean age 27 years, SD 5.1 years).

Results

EEG

The study population consisted of 17 patients (Table 1). Eight patients (47%) were investigated with depth EEG using both subdural and intracerebral electrodes because ictal EEG recording with conventional scalp electrodes as well as nasopharyngeal and/or sphenoidal electrodes did not permit a clear delineation of the area of onset of ictal EEG discharges. In all eight patients consistent unilateral epileptic foci were demonstrated with long term depth EEG-CCTV monitoring. In the other nine patients (53%) such invasive techniques had not yet been performed or were considered unnecessary since ictal and interictal EEGs with scalp electrodes and special electrodes (nasopharyngeal, sphenoidal electrodes) were considered sufficiently diagnostic. In four of these patients bilateral abnormalities were found demanding further invasive investigations. In the other five patients unilateral, ictal EEG discharges were found allowing for unequivocal lateralization of this activity but evidently not permitting localization of the epileptic focus with the same amount of precision as obtained in the patients studied with depth EEG recording.

CT and MRI

Neuroimaging with CTscan did not reveal any abnormality. In four patients (24%) MRI showed an increased signal intensity compatible with gliosis/sclerosis in the uncus/hippocampus area.

PET

PETscan showed a unilateral decrease of CMR Glu in two of four patients with bilateral scalp EEG abnormalities (Table 1). In another patient PET showed bilateral abnormalities and was normal in the fourth patient. There were 13 patients with unilateral EEG signs of epileptic abnormalitiy. In 12 of these 13 patients decreased glucose metabolism was found in more

Table 1. *Number of Patients Diagnosed with EEG and Showing Ipsilateral Abnormality with PET or SPECT*

EEG		(n)	PET	SPECT
Scalp	bilat.	4	2	3
	unilat.	5	4	5
Depth	unilat.	8	8	7*
		17	14	15*

* One patient showing contralateral abnormality.

or less wide spread areas. In all these patients the lateralization of PET abnormality, if present, correlated with the side of the epileptic focus as indicated by EEG data. When comparing localization by means of quantitative PET and depth EEG, disagreement between PET and EEG data was found in one patient. In this patient the maximal decrease of glucose uptake was found in the orbitofrontal region, whereas depth EEG showed a clearly circumscribed mesiotemporal onset of seizure activity. In the others the area of abnormality in PET coincided with, but exceeded the area of onset of ictal EEG discharges.

SPECT

In all patients interictal examinations were performed. In no patient did an alteration in the usual amount or severity of seizures occur. With visual assessment of SPECT, asymmetry was found in three of the four patients with bilateral scalp EEG abnormality (Table 1). Agreement of PET with SPECT data was found in only one of these patients. No further conclusions concerning correct or false lateralization could be drawn due to the lack of lateralizing EEG signs. In all five patients with unilateral seizure onset as found with scalp EEG, agreement concerning lateralization of EEG with SPECT data was present. In one of these patients PET showed no abnormality.

In six of eight patients studied with depth EEG the side of the decreased BZ receptor labelling coincided with the side of seizure onset in the ictal intracranial EEG recordings. In two other patients SPECT results were inferior to those of PET. In one patient no SPECT asymmetry was found. In the other patient a unilaterally decreased radioligand binding was shown contralateral to the side of onset of the ictal EEG discharges and of decreased glucose metabolism as shown with PET. In all 12 patients with unilateral epileptic foci the SPECT asymmetry involved widespread areas.

The spatial resolution of SPECT was insufficient to allow for a differentiation into frontal or temporal lobe abnormality.

Thus in one patient with a unilateral epileptic focus according to depth EEG recording PET showed abnormality in the hemisphere with the leading focus while SPECT erroneously indicated the contralateral hemisphere. In all other unilateral cases SPECT, when asymmetrical, was in agreement with the side of the epileptic focus as shown by EEG. PET, however, was normal in one of these cases. In all other 10 patients PET, SPECT and EEG were consistent. In the four patients (24%) with MRI signs of hippocampal gliosis agreement concerning lateralization existed between clinical, MRI, EEG, PET and SPECT data. The area of abnormality detected with PET and SPECT, however, clearly exceeded the hippocampal region.

Discussion

Quantitatively the inhibitory GABA-ergic system is the largest neurotransmitter system in the mammalian brain and is considered an important system of protection against seizures. GABA receptors form a complex with the Cl-ion channel in the neuronal membrane and GABA opens this channel after binding to its receptor. Subsequent influx of chloride ions leads to hyperpolarisation and thus to an inhibition of neuronal firing. BZ receptors are coupled to the GABA receptor/Cl^--ionophore complex, and BZ enhances the effect of GABA in an allosteric manner.

At present there are several radioligands with a high affinity for the BZ receptors. Flumazenil is a GABA neutral ligand, blocking the effects of both GABA-positive and GABA-negative ligands and having little or no sedative or anxiolytic effect of its own. By labelling this ligand with the positron-emitting isotope ^{11}C it has been made suitable for PET studies. Labelling a Flumazenil derivative with [123]I (^{123}I-Iomazenil) has been undertaken to produce a radioligand suitable for SPECT investigations.

A reduction of BZ binding has been shown in the cortex of kindled rats. Based on these animal observations, several authors have expected to find decreased BZ binding in cortical tissue excised from human epileptic patients. So far no clear alterations have been found[2,8]. However, in vivo studies with PET have demonstrated reduced ligand binding to BZ receptors in patients with complex partial seizures[7].

PET scanning has several drawbacks such as high cost and restricted availability. On the other hand it has relatively high spatial resolution and allows for quantification of data. So far PET scanning when using [18]FDG has been an important auxiliary method to identify localized metabolic abnormalities more or less congruent with epileptogenic area as assessed with EEG. One should realize, however, that [18]FDG PET reflects metabolic changes involving both neurons and glial cells. In contrast BZ receptors are restricted to the neuronal membranes and therefore more closely related to neuronal firing activity. Thus the demonstration of a decrease in BZ receptor labelling offers a new approach to the investigation of the epileptic focus per se. It has been claimed by Savic *et al.*[7], 1988, on the basis of data from one patient that the BZ receptor labelling technique delineates a smaller area of abnormality than does the [18]FDG PET technique, thus more closely corresponding to pathological and EEG findings.

The spatial resolution of the presently used PET was rather limited (in plane resolution 12 mm) and similar to that of SPECT. It is, however, a matter of debate whether the spatial resolution of the equipment used really matters, when studying metabolic processes since [18]FDG PET generally shows widespread functional abnormality. When applying quantitative analysis to PET data more precise results may be obtained. Presently, however, even with quantitative analysis the precision of PET is still insufficient to permit a reliable distinction between frontobasal, mesiotemporal and neocortical temporal epileptic foci.

As indicated in this study the area of reduced BZ receptor binding clearly exceeded the area of gliosis as demonstrated with MRI. Also the area of SPECT abnormality surpassed by far the area of onset of ictal EEG discharges as demonstrated with intracranial recordings to such an extent that a distinction between temporal and frontal lobe abnormality could not be made. Whether quantitative analysis of SPECT data or improvement of its spatial resolution will allow for a more precise localization requires further investigations.

Irrespective of its extent or cause the finding of decreased BZ receptor binding is in agreement with the hypothesis of inhibitory neurotransmitter (GABA) dysfunction in epilepsy.

In the present study in which ictal EEG was considered to provide reliable lateralization of the epileptic focus a high correlation between PET findings concerning glucose metabolism and SPECT data related to BZ binding was obtained in the 13 patients with unilateral epileptic foci, despite the completely different physiological parameters measured. In 10 patients complete agreement between results concerning lateralization was obtained. ^{123}I-Iomazenil SPECT showed false lateralization in one patient correctly diagnosed with [18]FDG PET. In two other patients either SPECT or PET provided correct lateralization while the other method showed no asymmetry. Thus ^{123}I-Iomazenil SPECT correctly identified the side of the epileptic focus in 11 of 13 patients (85%). This figure is far higher than those obtained with HMPAO-SPECT (59–67%) in other studies[1,6].

In conclusion ^{123}I-Iomazenil SPECT may be considered as a more economical and more widely available alternative for [18]FDG PET in the presurgical evaluation of patients with medically intractable complex partial seizures. Due to its limited spatial resolution, and/or the extent of abnormality concerned, it should be primarily used to identify the side of the epileptic focus, rather than to assess the exact localization or the extent of that focus. Both methods appear to be most reliable in patients with unilateral temporal seizure onset. The results obtained with these techniques are superior to those published recently for HMPAO-SPECT[1,6]. A theoretical advantage of ^{123}I-Iomazenil is that it specifically reflects changes in the functioning of membranes of neurons, whereas HMPAO reflects cerebral circulation and [18]FDG is related to glucose metabolism of both neurons and glial cells.

Acknowledgement

[123]I-Iomazenil was provided by the Paul Scherrer Institute (PSI), Würenlingen, Switzerland, in cooperation with Hoffmann-La Roche, Basel, Switzerland.

References

1. Andersen AR, Gram L, Kjaer L, Fuglsang-Frederiksen A, Herning M, Lassen NA, Dam M (1988) Spect in partial epilepsy: Identifying side of the focus. Acta Neurologica Scand [Suppl 117] 78: 90–95
2. Burnham WM, Hwang PA, Hoffman HJ, Becker LE, Murphy EG, Kish SJ (1987) Benzodiazepine receptor binding in human epileptogenic cortical tissue. In: Engel J Jr *et al* (eds) Fundamental mechanisms of human brain function. Raven Press, New York, pp 277–235
3. Engel J (1988) The role of neuroimaging in the surgical treatment of epilepsy. Acta Neurol Scand [Suppl 117] 78: 84–89
4. Franck G, Sadzot B, Salmon E, Depresseux JC, Grisar T, Peters JM, Guillaume M, Quaglia I, Delfiore G, Lamotte D (1986) Regional cerebral blood flow and metabolic rates in human focal epilepsy and status epilepticus. In: Delgado-Escueta AV *et al* (eds) Advances in neurology, Vol 44. Raven Press, New York, pp 935–948

5. Phelps ME, Huang SC, Hoffman EJ, Selin C, Sokoloff L, Kuhl DE (1979) Tomographic measurement of local cerebral glucose. Metabolic rate in humans with (F-18)2-fluoro-2-deoxy-d-glucose: validation of method. Ann Neurol 6: 371–388

6. Podreka I, Lang W, Suess E, Wimberger D, Steiner M, Gradner W, Zeitlhofer J, Pelzl G, Mamoli B, Deecke L (1988) Hexa-methyl-propylene-amine-oxime (HMPAO) Single photon emission computed tomography (SPECT) in epilepsy. Brain topography. J Funct Neurophysiol 1: 55–60

7. Savic I, Roland P, Sedvall G, Persson A, Pauli S, Widén L (1988) In-vivo demonstration of reduced benzodiazepine receptor binding in human epileptic foci. Lancet pp 863–866

8. Sherwin A, Matthew E, Blain M, Guévremont D (1986) Benzodiazepine receptor binding is not altered in human epileptogenic cortical foci. Neurology 36: 1380–1382

9. Theodore WH, Fishbein D, Dubinsky R (1988) Pattern of cerebral glucose metabolism in patients with partial seizures. Neurology 38: 1201–1206

Correspondence: A. C. van Huffelen, M.D., Department of Clinical Neurophysiology, State University Hospital, Heidelberglaan 100, 35849 Utrecht, The Netherlands.

Acta Neurochirurgica, Suppl. 50, 100–106 (1990)

Pre- and Postoperative Rehabilitation Related to Epilepsy Surgery

H. Silfvenius

Department of Neurosurgery, University Hospital, Umea, Sweden

Summary

A short review is presented of the mental impairment and personality deficits of persons with severe epilepsy and of their consequences, that require psychosocial and occupational rehabilitation, as well as of some aspects of health economics related to epilepsy surgery. It is concluded that detailed studies on the psychosocial outcome of epilepsy surgery are rather rare in the vast literature on epilepsy surgery, a situation that needs to be changed in order to promote better understanding of the therapeutic and palliative surgical treatment of epilepsy, to develop and broaden it, and to motivate more financial support from health authorities.

Keywords: Epilepsy; mental impairment; rehabilitation; surgical treatment; health economics.

Introduction to the Problem

This topic deals with the individual deficits of mental capability and personality definable by epileptological assessment, and with their consequences for educational, psychosocial and occupational insufficiencies which require rehabilitation. The primary goal of epilepsy surgery, in aiming at freedom from, or alleviation of seizures, is to assist in making a normal independent living possible for the treated individual. In order to achieve this goal, high quality epileptological assessment and care, as well as successful rehabilitation should be available and if necessary carried out. Much interest has so far correctly been focussed on the diagnostic and surgical aspects, on the outcome with respect to seizures and on the development of the treatment, while social rehabilitation has not received due attention. In 1958 this aspect was underlined by Green *et al.* stating that "the excision of the epileptogenic focus, avoidance of postoperative neurological deficit and achievement of economic independence and rehabilitation are the chief concerns and goals of therapy"[1]. The inadequate reporting on whether the patients were better off for their surgery has also been mentioned[2].

The scarcity of psychosocial studies related to epilepsy surgery has been further commented upon recently: "What changes in the patient's life does epilepsy surgery cause? There seems to have been a flight from the psychosocial field"[3].

Patients with severe partial epilepsy who are selected for epilepsy surgery have suffered from their disease usually for long periods of time before being operated on. Existing mental and personality deficits may be due not only to the structural lesion and the epileptic functional disturbances causing ictal periods of "social absence", but also to associated complications that demand extra care: wounds, burns, fractures, head injuries, job and traffic accidents, and forced absence from normal education, training and work. Such complications add to the negative influence of the disease on the educational, psychosocial, occupational and financial condition of the patient. Postictal states of confusion and amnesia are handicaps which disrupt the mental acitivity of the patients, leaving gaps in his/her contacts with daily life. Depending on the epileptic lesion and the additional disability, short or protracted care, counselling and rehabilitation may be necessary.

Interictal problems, intellectual and memory impairment, social shortcomings, anxiety – related emotional problems, low self-esteem, worries, concerns about self-fulfilment, social and interpersonal difficulties, marital status and especially unemployment/underemployment are common among persons with epilepsy[4]. Side-effects from antiepileptic drug treatment, periods of aberrant behavior, educational and sexual problems may become distressing experiences. Independent living and self support that would guarantee social integration and security is not always attainable. The causes of the interictal psychosocial problems may be subclinical, ictal EEG discharges or other types of EEG abnormalities[5,6]. Partial epilepsy has been shown to interfere with mental activity, a finding that also holds true for subclinical ictal EEG activity[7,8]. Only some 20% of patients with epilepsy seem to be without any significant intellectual, behavioural or neurological problems[9].

Ignorance of epilepsy and prejudice against it from other people and society in general may lead to depression or aggression in the patient and the relatives. The low level of public knowledge about epilepsy adds to the general discomfort[10]. Patients with temporal lobe seizures in particular were reported many years ago to be tolerated poorly by the community[11]. A study from England reported that 13% of the patients operated on for complex partial seizures were psychiatrically normal before operation and that 37% had family history of psychiatric disorder, 25% had disrupted school careers, and 37% were not usefully employed[12]. A Danish study of the psychosocial aspects of this patient group found that 17% were born out of wedlock and 73% were unmarried or divorced. They had a higher school age, and an inferior employment rate, about 44% before and after surgery, as compared to 66% in the general population[13]. The various ictal and interictal epileptic disturbances causing psychosocial and occupational problems are usually a feature of patients with drug resistant complex partial epilepsy of temporal or frontal lobe origin. These patients constitute about 2/3 of those offered epilepsy surgery[14]. Another important group in this respect are children with severe complex partial seizures and with large destructive epileptic lesions. If not treated early surgically, they become a heavy burden to themselves and their relatives and require protracted but rather unsuccessful education and rehabilitation[15,16]. Epileptological assessment of nonsurgical and surgical candidates is therefore the basis for subsequent, optimal psychosocial and occupational rehabilitation, the goal of which is to facilitate the patient's education and employment according to his skill, and to achieve an independent life. Reports in the literature give examples of how intensive anticonvulsant drug-monitoring and care of patients with severe epilepsy leads to improvement in seizure control and thereby optimizes psychosocial-occupational rehabilitation[17-19]. High quality epileptic diagnostic work up and care is best supplied within a specialized organization with different levels of competence and facilities. It is not unusual for a patient with epilepsy to be concerned about finding a physician interested in epilepsy, who has the competence to give adequate treatment. Referral stations like: Outpatient Epilepsy Units/Clinics, Specialized Neurological/Pediatric Departments, Epilepsy Centers and Comprehensive Epilepsy Centers are important and necessary links in such an organisation. At each level in this hierachy, educational, psychosocial-occupational counselling and help should be available.

Intensive monitoring of drug-treatment at an epilepsy unit preferably should include assessment by expert neuropsychologists to define the cognitive and emotional capabilites of the patient. This will define the basis for counselling and rehabilitation for educational, psychosocial and occupational needs.

Neuropsychological Testing for Educational, Psychosocial and Vocational Needs

Our knowledge of the psychosocial problems of epilepsy lags behind our medical understanding and care of the disability. In order to establish the mental capabilities on which psychosocial adjustment, education and job seeking should be based, neuropsychological screening should be carried out on children and adults having difficulties with their schooling or finding or maintaining a job.

Test batteries for adults assess the intellectual (e.g. WAIS, Wechlser Adult Intelligence Screening GATB, General Aptitude Test Battery) capacities of persons with deficits. Verbal and spatial perceptive impairments can be defined. An extensive neuropsychological battery for epilepsy may be used (MMPI-Minnesota Multiphasic Personality Inventory), which specifies verbal, spatial, memory, attentional, problem solving and other capacities. The WPSI (Washington Psychosocial Seizure Inventory) screens emotional problems and adjustment difficulties[20,21]. The pattern of psychosocial problems in epilepsy was recently studied in Seattle using the WPSI. It was established that only a handful of different problem-combinations account for the most prominent psychosocial problems, such as emotional-financial and emotional-vocational[22]. The Halstedt-Reitan test battery is best correlated with vocational adjustment and independent living, and can predict in 85% of adolescents if problems would arise later within these spheres[23,24]. Similar neuropsychological test batteries are also available for children.

The neuropsychological screening is important to the team responsible for the educational/vocational rehabilitation. This team may include a special teacher, a social worker, a family therapist, a vocational counselor, a psychologist, and a psychiatrist. Epilepsy disrupts training. Jobs are often lost because of emotional/

behavioral difficulties and various types of brain impairment[25, 26]. The emotional, psychosocial and occupational adjustments planned for the individual are related to the profiles obtained on testing and interviewing[22]. Emotional problems seem to be related to the presence of multiple seizure types[27, 28]. Interpersonal difficulties are experienced by some 50% of the patients with epilepsy[29]. The neuropsychological-psychosocial interrelationships have also been studied revealing that impaired IQ is associated with disordered adjustment[22]. About 75% of people with epilepsy have associated problems, difficulties, impairments, and handicaps. A detailed description of job counselling for people with epilepsy has been published[20].

Another important aspect of the problem is to quantify the psychosocial and occupational disabilities using functional health/disability scales. The deficits specified in daily life activities can be concerned with the subjective experiences of the disabilities and the shortcomings they lead to during leisure time, training and employment. Examples of quantifying health and disease with such scales are given in the social and the epilepsy literature[30–32]. Functional scaling is of particular importance if pre- and postoperative conditions are to be compared quantitatively, and even more important, if reports are to be made for administrators and politicians for health-economic reasons[33, 34].

Educational and Vocational Counselling

Education and social training of receptive children with epilepsy can be supported by early rehabilitation, underlining their privileges and commitments and avoiding overprotection from the parents. Special schooling and care arrangements are necessary for children with severe epilepsy. It has been reported from England, in a cohort study on children with epilepsy, that at the age of 11 years, 68% went to an ordinary school. When reinvestigated at the age of 16 years, 37% received special training in the school, 25% went to a special school and 7% to a school for children with epilepsy[35]. A German study reports special training in school in 42% of epileptic children[36].

It has been reported recently that 30% had lost their jobs because of epilepsy and that 25% were treated differently at their jobs because of their disease[37]. A study from the Netherlands showed the employment status in adult people with epilepsy to be 60% full time employed while 29% of those unemployed were unfit[38]. Unemployment for persons with epilepsy is reported to vary between 20–50%[26].

Vocational problems facing adults with epilepsy of various kinds are common. One study showed that about 50% had difficulties in securing a job. Discrimination in the job market was a serious problem. Job counselling was considered the service most needed and transportation to work was a major problem[39]. Recently it was established that about 60% of the adults with epilepsy reported concerns with their employment[29]. However, a person with epilepsy, when fit for the job, should be able to handle it like anyone else. The gaps in his mental and social capacities and skills due to the illness should be minimized by special training at work, aiming to counteract the negative effects of the disease[40]. Neuropsychological differences have been found between employed and unemployed persons with epilepsy, particularly with regard to those pertaining to memory, alertness and cognitive flexibility. Recent investigations on the employment of persons with epilepsy in general and some surgically treated in particular have been published. They have shown that of those with generalized seizures (not operated cases) and of working age, 46% were employed, 14% unemployed, 18% retired, the remaining 22% were students and housewives and 55% had a handicap status[37].

General Remarks on Psychosocial Rehabilitation

The results of attempts to alleviate or cure the psychosocial problems of epilepsy have been poorly quantified and described in the literature on epilepsy surgery. Rehabilitation includes training in social and independent living skills, educational assessment and goal setting, family counselling, job counselling, and job and community follow up and each such important aspect must be critically evaluated. Such rehabilitation can be assisted by Self-Help groups and training courses[41–43]. Reports on measures taken by national epilepsy societies and authorities to encourage, improve and support social integration and employability have been presented. In the USA, TAPS, (Training and Placement Service Project), in the United Kingdom, DRO, (Disablement Resettlement Officer, a government funded post), and in Germany, Vocational Training Center, participate in the social reconstruction and integration of the patients[44–46]. Good results for employability and client satisfaction is reported from counselling services[47–49].

Postoperative Psychosocial, Educational and Occupational Rehabilitation

Once the partial epilepsy has been found to be drug resistant, the possibility of surgical treatment should be discussed. If carried out, it might be of interest considering the accumulated findings preoperatively, to classify the case as therapeutic, with a good chance of becoming a surgical success, or as palliative with less expectation of seizure relief. Such a committed classification coerces the medical professionals to build up a detailed assessment program and to devise a postoperative psychosocial rehabilitation program. That program could be divided into therapeutic and palliative parts, both having their own goals and outcome scales. This could help in establishing predictive psychosocial, educational and occupational tests, that are not reliant on neuropsychological screening methods alone, but that also consider the effects of surgery on the care load of society.

Nowadays only occasionally are the psychosocial and occupational aspects specified in detail in the preoperative assessment. Some studies carried out recently emphasize, however, such pre- and postoperative psychosocial and occupational findings.

At the Maudsley Hospital, London, one of the forerunners in this field, studies have been carried out on the effects of epilepsy surgery for complex partial seizures on the psychiatric symptoms of the patients. It was found that only those with preexisting personality disorders benefited from surgery[50]. Another report mentioned that 72% had a character disorder, and that the most apparent result of the surgery was reduced aggressiveness. The patient's social relationships also improved. Socioeconomic and psychological improvement was reported to become normal in 17% after the operation. Improvements were significant in work and human relationships within and beyond the family. It was found, however, in a follow up, that 37% had deteriorated mentally and that 5% had committed suicide. Taylor, however, states that "it is not universal, necessary or irremedial that mental and social disorder is associated with temporal lobe epilepsy"[3]. A Danish study on the results of surgery for complex partial seizures reports 28% of the patients institutionalized preoperatively, and 8% postoperatively. They had a lower social class than their parents, which was lower than that of the general population[13]. This material presumably is highly selected and presents a rather depressing picture of a population of complex partial epilepsy. A series on surgery for complex partial epilepsy from the USA reports worthwhile results overall in the majority of the cases. The non-improvement depended on cognitive impairment and defective personality, development, depression and paranoia[51]. More recent result of the beneficial effect of epilepsy

surgery on the disturbed personality have been reported in other series[15, 16, 50, 52, 53].

Postoperatively transient psychosocial problems may appear in about 30–40% of those operated on for focal epilepsy, often of a reactive tpye[54]. The pre- and postoperative profiles of psychosocial functioning may improve or be impaired depending on the aspects studied. The best outcome from surgery is seen among those who preoperatively are "normal"[55]. Some studies report that of those living in an institution preoperatively, postoperatively 47% could manage to live in a home, but also that only 13% of those operated on for complex partial epilepsy could live independently[55]. A recent report underlines, however, that patients operated for complex partial epilepsy improve in their independence, working capacity and external social contacts[50].

It would be of great interest to specify in detail the reasons for a person with epilepsy being institutionalized. The roles played by insufficiencies in social setting, upbringing, education, family and society might have been of equal or even greater importance than the disease itself in leading to a sheltered life.

A well controlled recent study specifically focussed on the outcome of surgery for complex partial epilepsy on employment and psychosocial functioning, showed that employment increased from 44% to 72%, underemployment decreased from 25% to 0%, but that the percentage of those who were unemployed did not change by surgery[56]. A major drawback is that epilepsy surgery is generally carried out at a relatively late stage when the patient voluntarily or otherwise has already become accustomed to a more or less sheltered existence.

It is particularly important to offer pediatric epilepsy surgery early. Its beneficial psychosocial effects have been documented in several reports both on complex partial epilepsy and on epilepsy due to large destructive lesions. If not carried out, protracted psychosocial and occupational misery is waiting ahead[57–61].

Gross structural lesions lead to a major handicap and an advanced behaviour disorder. Rage attacks and profound aggression against family may be a daily heavy burden. Schooling is disordered by fits, and disruptive behaviour often stops further schooling. There is progressive deterioration of IQ. The social outcome after hemispherectomy in 17 children for large destructive lesions, which led to freedom from seizures, was an immediate change in behaviour with disappearance of aggression and the children became friendly and cheerful. Only a minority required residential care for social reasons. About half of these treated in the appropriate age group were employed in simple work. All children of school age were in fulltime education, in normal or in handicap schools. No patient had suffered any detectable IQ or social deterioration[15].

Twenty-nine children operated on for complex partial epilepsy were studied pre- and postoperatively with regard to education, employment and social adjustment. About 1/3 had been in normal schools and returned to them and the other 2/3 had been excluded. Of these, 1/3 achieved normal schools, 1/2 attended special schools, and a few were unsuitable for education. The patients operated on after the age of 21 years were in employment or house-wives. Many of the 27 in the school age group attended remedial classes in reading and mathematics or entered fulltime education or training. Improvement in social behaviour and psychiatric state were notable, though protracted rehabilitation was needed in many cases[16]. A study from the USA on children operated on for complex partial seizures also reported social improvement and a large portion of these patients were either in school or actively employed[62].

The positive effects of epilepsy surgery on educational receptiveness, psychosocial and vocational functioning, should receive greater attention than hitherto. This would imply that the treatment team emphasizes these aspects, studies them carefully pre- and postoperatively and presents their results quantitatively. This requires the use of functional scales which express the findings in a measurable way to the patient, treatment team, administrators and politicians[63]. The latter might in turn become more convinced about the beneficial effects of epilepsy surgery in making the majority of the patients seizure free or markedly reduce their seizures at very low risk and costs, and change the psychosocial situation for the better. Thereby more resources could be re-allocated for epilepsy surgery.

Health Economies and Epilepsy Surgery

The costs of health care may be expressed as direct and indirect expenditures. The direct costs arise partly in the professional health sector: prevention, diagnosis, hospital payments, drugs and similar costs. The indirect costs are to a large extent due to losses in production by morbidity and mortality. The direct costs are far less than the indirect ones and therefore there seems to be no health economic reasons against increased treatment with epilepsy surgery. Details of health economics as related to epilepsy and epilepsy surgery can

be found in the literature[33, 34, 63, 64]. Recently the indirect costs for mortality from epilepsy during 1971–1984 of those deceased in production age were estimated for Sweden to be about 300 million crowns ($ 50 million). The indirect costs due to loss of production because of epilepsy in the corresponding age group was calculated on the basis of the number of persons receiving disability pensions. These indirect costs, which are transfer costs, also amounted to about 300 million Swedish crowns. The actual cost for the disability payments from the state because of epilepsy during 1971–1984 was about 2.4 billion crowns ($ 400 million). An extrapolation of the costs of the loss of future production of those receiving these disability pensions with reference to the average life expectancy was estimated to be about 6.500 million crowns ($ 1.000 million)[63]. Such magnitudes of indirect costs should be appreciated and considered when the relatively small direct costs for epilepsy surgery are mentioned. It is helpful to separate the costs and gains of therapeutic and palliative surgical treatment (excisions, callosotomies). The economic gain from even partial improvement after palliative surgery is the consequence of the reduced need for supervised care by society. The cost-benefit outcome for establishing a number of Epilepsy Clinics/Units in Germany has been reported. The study specified the gain in functional level of the treated patients and the large cost reduction for the state from this suggested organization[32]. Cost benefit results have also been presented from Australia underlining the savings in social costs if drug-treatment is carried out by expert epileptologists[33]. The potential savings in the USA from prevention and care of epilepsy were specified in a comprehensive state report in 1978[64].

References

1. Green JR, Steelman HF, Duisberg RE *et al* (1958) Behaviour changes following radical temporal excision in the treatment of focal epilepsy. Proc Assoc Res Nerv Ment Dis 36: 295–315
2. Bates JAV (1962) The surgery of epilepsy. In: Williams D (ed) Modern trends in neurology, vol 3. Butterworths, London
3. Taylor DC (1987) Psychiatric and social issues in measuring the input and output of epilepsy surgery. In: Engel J Jr (ed) Surgical treatment of the epilepsies. Raven Press, New York, pp 485–503
4. Collings JA (1987) The social, psychological, and physical well-being of people with epilepsy in the community. XVIIth Epilepsy Internatl Congr. Jerusalem, 23
5. Dodrill CB, Wilkus RJ (1978) Neuropsychological correlates of the electroencephalogram in epileptics. III. Generalized non-epileptiform abnormalities. Epilepsia 19: 453–462
6. Zivin L, Ajmone-Marshan C (1968) Incidence and prognostic significance of "epileptiform" activity in the EEG on non-epileptic subjects. Brain 91: 751–778

7. Loiseau P, Strube E, Broustet D *et al* (1980) Evaluation of memory function in a population of epileptic patients and matched controls. Acta Neurol Scand [Suppl 80] 62: 58–61

8. Aarts JHP, Binnie CD, Smith AM *et al* (1984) Selective cognitive impairment during focal and generalized epileptiform EEG-activity. Brain 107: 293–308

9. Rodin EA, Shapiro HS, Lennox K (1976) Epilepsy and life performance. Lafayette Clinic, Detroit

10. Petrella MA *et al* (1987) Level of public information on epilepsy: An epidemiological investigation in Italy. XVIIth Epilepsy Internatl Congr Jerusal, 94

11. Bailey P, Gibbs FA (1951) The surgical treatment of psychomotor epilepsy. JAMA 145: 365–370

12. Taylor DC (1972) Mental state and temporal lobe epilepsy. A correlative account of 100 patients treated surgically. Epilepsia 13: 727–765

13. Jensen I (1972) Social conditions of temporal lobe epileptics in Denmark, Epilepsia 13: 71–74

14. Engel J Jr (1987) Outcome with respect to seizures. In: Engel J Jr (ed) Surgical treatment of the epilepsies. Raven Press, New York, pp 553–572

15. Lindsay J. Ounsted C, Richards P (1987) Hemispherectomy for childhood epilepsy: A 36 year study. Dev Med Child Neurol 29: 592–600

16. Lindsay J, Ounsted C, Richards P 81984) Long-term outcome in children with temporal lobe seizures. V. Indications and contraindications for neurosurgery. Dev Med Child Neurol 26; 25–32

17. Janz D (1988) Neurological morbidity of severe epilepsy. Epilepsia 29 [Suppl 1]: S 1–8

18. Sutula TP, Sachellares JC, Miller J *et al* (1983) Intensive monitoring in refractory epilepsy. Neurology 431: 243–247

19. Theodore WH, Schulman EA, Porter RJ (1983) Intractable seizures: Long term follow-up after prolonged inpatient treatment in an epilepsy unit. Epilepsia 24: 336–343

20. Fraser RT (1984) Improving the employability of those with epilepsy. In: Porter RJ *et al* (eds) XVth Epilepsy Internatl Symp, Adv Epileptol, pp 599–604

21. Dodrill CB (1988) Neuropsychology. In: Laidlaw J, Richens A, Oxley J (eds) A textbook of epilepsy. Churchill Livingstone, London, pp 406–420

22. Dodrill CB (1986) Psychosocial consequences of epilepsy. In: Filskov SB, Poll TJ (eds) Handbook of clinical neuropsychology, Vol 2, Chapt 13. Wiley, New York, pp 338–363

23. Heaton RK, Pendleton MG (1981) Use of neuropsychological tests to predict adult patients' everyday functioning. J Consult Clin Psychol 49: 807–821

24. Clemmonds DC, Dodrill CB (1984) Vocational outcomes of high school students with epilepsy and neuropsychological correlates with later vocational success. In: Porter RJ *et al* (eds) XVth Epilepsy Internatl Symp, Adv Epileptol. Raven Press, New York, pp 611–614

25. Dodrill CF (1987) Neuropsychological assessment – does it help in job placement and retention? XVIIth Epilepsy Internatl Congr. Jerusalem, 34

26. Batzel LW, Dodrill CB, Fraser RT (1980) Further validation of the WPSI vocational scale: comparisons with other correlates of employment in epilepsy. Epilepsia 21 (3): 235–242

27. Herman BP, Dikmen S, Wilensky AJ (1982) Increased psychopathology associated with multiple seizure types: Fact or artifact? Epilepsia 23: 587–596

28. Rodin EA, Katz M, Lennox K (1976) Differences between patients with temporal lobe seizures and those with other forms of epileptic attacks. Epilepsia 17: 313–320

29. Dodrill CB (1983) Psychosocial characteristics of epileptic patients. Res Publ Assoc Res New Ment Dis 61: 341–353

30. Berdit M, Williamson JW (1973) Function limitation scale for measuring health outcome. In: Berg RL (ed) Health status indexes, Hospital Research Trust, Chicago, 57

31. Patrick DL, Bush J, Chen M (1973) Toward an operational definition of health. J Health Soc Behav 14: 6–23

32. Thorbecke R, Janz D (1984) Guidelines for assessing the occupational possibilities of persons with epilepsy. In: Porter RJ *et al* (eds) XVth Epilepsy Internatl Symp, Adv Epileptol. Raven Press, New York, pp 571–576

33. Ionnaides-Demos L, Horne M, McNeil J *et al* (1987) Cost effective analysis of a pharmacokinetic/therapeutic intervention service in an epilepsy clinic. Clin Pharmacol Ther 41: 214

34. Kriedel T (1980) Cost-benefit analysis of epilepsy clinics. Soc Sci Med 14: 35–39

35. Ross EM, Kurtz Z, Peckham CS (1983) Children with epilepsy: Implications for School Health Service. Publ Hlth, London, 97: 75–81

36. Doose H, Sitepu B (1983) Childhood epilepsy in a German city. Neuropediatrics 14: 220–224

37. in der Beek R, Mörchen G, Ritter G (1987) Employment status, occupational qualification and occupational fields of epileptics in an area serviced by a neurological clinic. XVIIth Epilepsy Internatl Congr, Jerusalem, 81

38. Rutgers MJ (1987) Employment status in adult people with epilepsy. XVIIth Epilepsy Internatl Congr, Jerusalem, 102

39. Perlman LG (1977) The person with epilepsy: Lifestyle, needs, expectations. National Epilepsy League, Chicago

40. de Boer HM (1987) The component of a good vocational program. XVIIth Epilepsy Internatl Congr, Jerusalem, 31

41. Fraser RT, Smith WR (1982) Adjustment to daily living. In: Sands H (ed) Epilepsy: A handbook for the mental health profession. Brunner/Mazel, New York, pp 189–221

42. McGovern S, Jones OM (1980) The development of self-groups. In: Wada JA, Penry JK (eds) Xth Epilepsy International Symposium, Adv Epileptol. Raven Press, New York, pp 413–416

43. Engel J Jr (1988) Surgical treatment of epilepsy. In discussion. Acta Neurol Scand [Suppl 117]: 147–149

44. Dellinger R (1984) Keynote address. In: Porter RJ *et al* (eds) XVth Epilepsy Internatl Symp, Adv Epileptol. Raven Press, New York, 567

45. Aspinall A (1984) Employability of the person with epilepsy. In: Porter RJ *et al* (eds) XVth Epilepsy Internatl Symp, Adv Epileptol. Raven Press, New York, pp 587–588

46. Tynova L (1987) Aspects of vocational rehabilitaion of epileptics at the Heidelberg vocational training centre. XVIIth Epilepsy Internatl Congr, Jerusalem, 127

47. Peterson HD (1987) Overcoming the problem of industry regarding employment issues and epilepsy. XVIIth Epilepsy Internatl Congr, Jerusalem, 16

48. Hayes CV (1984) Factors influencing employability: An analysis of the successful training and placement service. In: Porter RJ *et al* (eds) XVth Epilepsy Internatl Symp, Adv Epileptol. Raven Press, New York, pp 593–598

49. Gorsky D (1987) Marketing approach to job placement. XVIIth Epilepsy Internatl Congr, Jerusalem, 50

50. Rausch R, Crandall PH (1983) Psychological status related to surgical control of temporal lobe seizure. Epilepsia 23: 191–202

51. Horowitz MJ, Cohen FM (1968) Temporal lobe epilepsy, effect of lobectomy on psychosocial functioning. Epilepsia 9: 17–22

52. Jensen I, Larsen K (1979) Mental aspects of temporal lobe epilepsy. J Neurol Neurosurg Psychiatry 42: 256–265

53. Polkey CE (1983) Effects of anterior temporal lobectomy apart from the relief of seizures: A study of 40 patients. J R Soc Med 76, 354–358

54. Hurley O et al (1983) Life and emotional status in the first year following temporal lobe resection. 15th Epilepsy Internatl Symp, Washington DC

55. Fraser A (1988) Improving functional rehabilitation outcome following epilepsy surgery. Acta Neurol Scand [Suppl 117] 78: 122–128

56. Augustine EA, Novelly RA, Mattsson RH et al (1984) Occupational adjustment following surgical treatment of epilepsy. Ann Neurol 15/1: 68–72

57. Blume WT, Girvin JP, Kaufmann JCE (1982) Childhood brain tumours presenting as chronic uncontrolled focal seizure disorders. Ann Neurol 12: 538–541

58. Davidson S, Falconer M (1975) Temporal lobectomy in children with epilepsy. Lancet i: 1260–1269

59. Drake J, Hoffman, Kobayshi J et al (1987) Surgical management of children with temporal lobe epilepsy and mass lesions. Neurosurgery 21/6: 792–797

60. Goldring S (1987) Surgical management of epilepsy in children. In: Engel J Jr (ed) Surgical treatment of the epilepsies. Raven Press, New York, pp 445–464

61. Rasmussen T (1983) Cortical resection in children with focal epilepsy. In: Parsonage M et al (eds) XIVth Internatl Epilepsy Symp, Adv Epileptol. Raven Press, New York, pp 249–254

62. Meyer FB, Marsh WR, Laws ER Jr et al (1986) Temporal lobectomy in children with epilepsy. J Neurosurg 64: 371–376

63. Silfvenius H (1988) Economic costs of epilepsy – treatment benefits. Acta Neurol Scand [Suppl 117] 78: 136–144

64. Plan for Nationwide Action on Epilepsy (1978) Department of Health, Education and Welfare, National Institutes of Health, DHEW Publication No (NIH) 79–1115, Washington

Correspondence: H. Silfvenius, M.D., Ph.D., Department of Neurosurgery, University Hospital, S-901 85 Umea, Sweden.

Acta Neurochirurgica, Suppl. 50, 107–116 (1990)

Why Are so Few Patients Operated on for Epilepsy? A Yugoslavian Perspective

I. I. Ribarić

Neurosurgical Clinic, University Clinical Centre, Belgrade, Yugoslavia

Summary

A detailed review of the literature is provided which documents the discrepancy between the number of potential candidates for epilepsy surgery and the number actually operated on. Against a historical background, the limitations of understanding and of diagnostic tests are evaluated together with a careful review of long term surgical results, including those of the author.

Keywords: Epilepsy; surgical treatment; preconditions; results.

I. There Is a Big Discrepancy Between the Number of Candidates for Epilepsy-Surgery (C) and the Number of Patients Operated on (O) – ("C-O discrepancy")

Some recent reports indicate a prevalence rate of focal seizures of 4.0 per 1,000 population [27, 58]. It is reported that there are 0.49/1,000 new cases of epilepsy per year [85] of which 50–75% suffer from focal seizures [27, 85]. Approximately 45% of patients with active partial seizures remain poorly controlled by medication [27, 61]. Hence using modern criteria for surgery, it may be estimated that 12.5–25% of persons with drug-resistant partial epilepsy may be candidates for neurosurgical intervention [9].

Applying these statistics to the United States population it has been estimated that using modern diagnostic techniques, including positron emission tomographic scanning (PET), stereo-electroencephalography (SEEG) and intensive monitoring with radio telemetry-recorded ictal events, 75,000 persons might be eligible for surgical treatment by traditional criteria [4], and there are an additional 4,000 new candidates per year. Barely 300 of these procedures are performed in the USA annually [13].

If we apply these statistics to the Yugoslav population, we can calculate that about 8,000 persons are candidates for surgical treatment of medically intractable epilepsy and about 500 additional new candidates emerge annually. I have operated on 300 patients since 1974, and I have been the only neurosurgeon performing these kinds of operations in Yugoslavia.

A similar C-O discrepancy exists in most other countries in the world.

These numbers indicate that epilepsy-surgery has been underdeveloped and kept peripheral to the mainstream of neurosurgery. By way of illustration of this statement is the fact that among the main topics at the past eight European Congresses of Neurosurgery (Zürich, 1959; Rome, 1963; Madrid, 1967; Prague, 1971; Oxford, 1975; Paris, 1979; Brussels, 1983; Barcelona, 1987) there was none devoted to the surgical treatment of epilepsy.

II. Is the Present C-O Discrepancy Understandable?

A. A Brief Insight into the History of Epilepsy-Surgery

In the period *between 1860 and 1880*, the first scientific approaches to the mystery of the brain and the mind were made: Paul Broca, a French surgeon and pathologist, showed in 1870 [3] that speech function has a localization within the brain of man; Fritsch and Hitzig, German physiologists, in 1861 [20] proved by electrical stimulation that motor function can be localized within the precentral gyrus of a dog; John Hughlings Jackson, English neurologist, listening to the patients who described to him their epileptic seizures or watching a fit at the bedside, surmised in 1873 [74] that this was evidence of brain action caused by unruly discharge in some area of grey matter.

In 1886, Sir Victor Horsley performed the first operation for epilepsy [28].

Between 1920 and 1940, the basis for the present technique of resective surgery for epilepsy was established. The technique of cortical resection stems from

Otfrid Foerster's operations in the 1920's[17], but it was Wilder Penfield's scholarly and scientific work in the Montreal Neurological Institute, stimulated by his studies with Foerster[19] and accompanied by Herbert Jasper's electroencephalographic genius, that have largly been responsible for the development of cortical resection as a valuable therapeutic technique for a selected group of patients with focal seizures[50]. Electroencephalography, the crucial diagnostic methodology for demonstrating that a patient's epileptic seizure originate in a circumscribed area of the brain was founded by Hans Berger in 1929[2] and was mandatory for the development of epilepsy-surgery. It was very quickly applied to the field of epilepsy by Frederick and Erna Gibbs and William Lennox in USA, Henry Gastaut in France, and particularly by Herbert Jasper, who joined in 1939 the staff of the Montreal Neurological Institute.

Between 1950 and 1960, many valuable achievements emerged. In 1953, a unique "temporal" clinical partial seizure type[21] was established, its associated EEG abnormalities[21, 63] and an associated, predominantly hippocampal pathology, including the "theory" of incisural sclerosis or perinatal tentorial compression as the cause of hippocampal sclerosis. Sphenoidal electrodes, although first used by Kristensen and Rey[31] were improved in 1960 by Rovit, Gloor, and Henderson[62]. Electrocorticography, although introduced by Foerster and Altenburger in 1935[18], was developed by a number of authors in the late 1950's[32, 45, 79]. The technique of implantation of chronic depth electrodes into selected areas of the brain flourished after the publication of atlases providing their spatial coordinates by Talairach *et al.* in 1957[72, 73] and Schaltenbrand and Bailey in 1959[64]. The studies of intracerebral electrical activity in epileptic patients have been performed by a great number of clinicians[80] contributing enormously to our understanding of the origin and spread of seizure discharges.

Between 1970 and 1980, diagnostic techniques emerged which contribute very much to the preoperative evaluation of the candidates for epilepsy-surgery: long-term telemetric EEG recording was developed in the late 1970s[49, 69, 70]: Although brain imaging was initiated by ventriculography and pneumoencaphalography introduced by Walter Dandy in 1918 and 1919[5, 6], and greatly improved with angiography introduced by Egaz Moniz in 1927[39], it was not until 50 years later that computerized tomography (CT), introduced by Hounsfield in 1973[29], made it possible to visualize the smallest brain lesions which frequently accompany epileptic discharges. More recently still has seen utiliza-

tion of Nuclear Magnetic Resonance for brain imaging[30, 33] (MRI), as well as Positron Emission Tomography (PET)[15].

The preceding historical insight shows that the diagnostic procedures essential for the contemporary resection technique in epilespy-surgery became clinically available in the last 1–3 decades. Our systematic clinical understanding of epilepsy has advanced in parallel. Although classifications of seizures had been previously attempted, the first to be internationally accepted was the classification published by Gastaut in 1970[22] and the most recent classification of epilepsies and epileptic syndromes was published in 1985[10].

B. Theoretical Gaps and Insufficiency of Diagnostic Tests

1. In spite of important new discoveries in the neurosciences and molecular genetics in the last two decades[8], there remain many unanswered questions about the cellular and molecular mechanisms by which epileptic attacks inherently start, regenerate, arrest spontaneously, or progress to status epilepticus, mechanisms of their spread and their induction of chronic enhanced states of excitability, and eventually production of cell damage. At present there is agreement that seizure activity runs in parallel with characteristic, steep, high-amplitude potentials in the EEG, which indicate a synchronous abnormal activity of a large number of neurons in the area surrounding the electrodes. Intracellular recordings during focal seizure activity revealed that the cellular equivalent of the seizure discharges is the unique behaviour of the neurons involved in the epileptic process, which was named "paroxysmal depolarization shift" (PDS)[26, 37]. The majority of experimental data are in favour of the concept that PDS is the result of altered intrinsic properties of the neurones directly involved in the epileptic process[11]. The question is which factors are crucial for the alteration of the intrinsic properties of the neurones responsible for their unique epileptic behaviour? One of the attractive explanations remains that progressive deafferentation contributes to neuronal epileptogenicity through release of inhibitory input. But it is well known that the epileptic focus cannot be defined by a single pathognomonic morphological change. It rather reflects the effects of various structural and molecular abnormalities on selectively vulnerable neuronal clusters and supporting cells.

2. Clinical, surgical, histological, electrophysiological, and biochemical data suggest that the epilepto-

genic area is often either in the region of the brain tissue pathology (focal cortical dysplasia, neuronal loss or gliosis of the cortex or white matter, Ammon's horn sclerosis, mesial temporal sclerosis[1], macrogyria[75], forme fruste of tuberous sclerosis[48], chronic encephalitis of Rasmussen type[53], infarcts or scars) or in the close vicinity of "foreign tissue" lesions (calcified angiomata, cavernous hemangiomata, arteriovenous malformations, calcified parasitic granulomata, Sturge-Weber pathology, traumatic brain lesions, tumors, heterotopias[83], hamartomas, gangliogliomas, tubers, arachnoid cysts). Surgical results with respect to seizure control are best when the epileptic focus is spatially related to some of the aforementioned lesions. However, each of the aforementioned brain or "foreign tissue" lesions may not be accompanied by epilepsy or the epileptogenic area may be located at a distance from the lesion visualized by some of the brain imaging techniques.

In my series of 224 patients operated on for drug-resistent "temporal" epilepsy I found at operation very often some anatomical abnormalities which had not been emphasized in the literature referring to epilepsy, but their eventual contribution to development of an epileptogenic focus in "temporal" epilepsy should be examined. Such anatomical abnormalities are:

(a) tentorial herniation of the uncus, which reaches sometimes more than 2.5 cm below the tentorial edge, and occasionally is situated medially to the posterior cerebral artery. Due to the great frequency of the uncal infratentorial position, this structure escapes examination with depth cerebral electrodes. The uncal contribution to the epileptic properties of mesial temporal structures is not clarified.

(b) at the base of the middle cranial fossa I noticed many times one or more small (a few mm) indentations of brain and meninges into the shallow bone excavations, which I call "microencephalomeningocoeles". On a few occasions I found sharp bone spicules impressed into the inferior temporal cortex.

(c) in my series of 74 patients operated on with my technique of "restricted temporo-frontal resection"[59] I found very often two anatomical abnormalities: "posterior sphenoidal herniation" with the impression of the sphenoidal ridge into the posterior frontoorbital cortex, and spicule-like or irregular bone prominences elevated up to two centimeters above surrounding bone, with corresponding impressions in the frontoorbital cortex.

In contrast, there are brain areas identified with EEG and clinically as epileptogenic foci, without de-

monstrable morphological lesions and prone to age-related spontaneous remission. At present we know two such epileptic syndromes, called "Idiopathic localization-related epilepsies": benign childhood epilepsy with centrotemporal spikes and childhood epilepsy with occipital paroxysms[10]. Which factors produce "epileptogenic neurons" in these cases? Which factors contribute to reversibility of neuronal epileptic behaviour? Which factors are essential for the influence of brain maturation on the brain's epileptogenic properties (in many cases seizure onset is age dependent; epileptic foci and habitual seizure properties are changing during brain maturation; there is a tendency for epileptogenic foci to migrate during brain maturation from the posterior towards anterior brain structures)?

3. None of the diagnostic tests presently included in presurgical evaluation protocols are absolutely reliable in the localization of an epileptogenic brain area, but the reliability increases progressively with the number of tests with congruent results.

a) *Radiographic studies*, in particular computed tomography (CT) and magnetic resonance imaging (MRI) have become important tools for the localization of structural abnormalities in patients with epilepsy, although visualization of structural abnormalities may not necessarily indicate the site of the epileptogenic area. MRI has a greater spatial resolution than CT, but it does not image calcium and can miss calcified lesions seen by CT, and it does not appear to be capable of identifying mesial temporal sclerosis, the most common lesion encountered in temporal lobe epilepsy[68].

Positron emission tomography (PET) offers the opportunity to image brain metabolism and blood flow. In epilepsy, most PET studies have measured glucose metabolism, employing deoxyglucose labeled with fluorine-18 (FDG). Interictal focal hypometabolism was found in about 70% of patients with temporal epilepsy[15, 76]. A high correlation was noted between the site of the hypometabolic region and the site of ictal onset, recorded with scalp or depth EEG, but Engel and colleagues[15] noticed that there is no difference in the outcome with respect to seizures following surgery between the 44 patients who had positive findings on PET scan and 21 patients who had normal PET scans.

Single-photon emission computed tomography (SPECT) was initially developed to provide images of regional cerebral blood flow (rCBF). This technique has limited sensitivity and studies in epilepsy have not yet been rigorously correlated with EEG, pathological examination and other diagnostic tests used for determining localization of epileptogenic area[67].

b) *Neuropsychological evaluation* of the epileptic patients considered for surgical treatment has been used for diagnostic and predictive purposes. Patients with a functional deficit restricted to the temporal lobe of intended resection are the most likely to benefit from surgery[55], and the amount of cognitive loss following a standard temporal lobe excision is directly related to the level of preoperative cognitive performance. The value of neuropsychology in predicting seizure control following extratemporal lobe surgery has not been systematically investigated. The rationale for using neuropsychological evaluation to detect focal brain dysfunction is based on the close correlation between neuropsychological deficit and definitive pathological or surgical lesions. However, the ability of neuropsychological tests to detect epileptic tissue that may not be grossly damaged has not been well explored[54] and one should bear in mind that epileptic focus may disrupt behaviour in a different manner to a structural lesion.

c) *The early symptoms and signs* of the patient's spontaneous habitual ictal episodes, as well as ictal and postictal behaviour, remain an invaluable aid in the localization of a possible epileptogenic brain areas. Our current understanding of anatomico-clinical correlations are based mainly on the reports from the extensive surgical programmes of the Montreal and Paris schools directed either at electrostimulation or recording with acutely and/or chronically implanted depth electrodes. The important steps in completing description of ictal clinical manifestations are medical records, seizure history from patient and from witnesses, and closed-circuit television monitoring combined with simultaneous EEG recording (CCTV/EEG), preferably with interactive examination of the patient during the ictal episodes. The localizing value of individual auras, motor patterns, or automatism should never be considered as crucial, but simply suggestive. Certain features are more suggestive than others.

d) *Electrographic recordings* of epileptoform abnormalities are most important for identification of epileptogenic brain area, because they reflect most directly the presence of epileptogenic neuronal hyperexcitability in the form of interictal or ictal epileptiform discharges[24]. In spite of that, there are numerous problems influencing the interpretation of electrographic data. Epileptiform abnormalities, even if localized, may reflect abnormal activity originating from a distant site, not directly accessible to EEG recording, a factor which presents a problem in both extracranial and intracranial EEG recordings. EEG potentials can be recorded over a considerable distance from the area that generates them by virtue of volume conduction. The location, polarity, field configuration, and even recordability with extracranial electrodes of a potential generated by a cortical generator is dictated not only by its location in relation to the recording electrodes, but also by its size, its geometric orientation, its configuration and by the electrical inhomogeneities of tissues placed between the cerebral generators and the extracranial electrodes. Scalp EEG recording has been improved by the advent of three major contributions.

(1) The use of special basal electrodes, such as true temporal, earlobe, nasopharyngeal, sphenoidal, supraorbital, and orbital and foramen ovale to localize EEG abnormalities to mesial and ventral brain areas. Recent evidence suggests nasopharyngeal electrodes contribute little over earlobe or true temporal derivations to justify the additional discomfort, whereas sphenoidal electrodes do improve localization and increase the yield of positive tests[66].

(2) The development of direct and telemetric long term recording techniques in order to determine EEG correlates of spontaneous seizures.

(3) Computer approaches to the acquisition, storage, analysis and display of EEG data to maximize the ability to identify meaningful abnormalities.

There is controversy concerning the reliability of localization of an epileptogenic lesion on the basis of interictal EEG spikes alone. Certainly ictal EEG and other studies are useful when the interictal EEG is equivocal, but whether additional information derived from ictal EEG is of critical importance in preventing an inappropriate surgical resection has not been established. It is possible that ictal data on localization are no more accurate than interictal data[16]. However, confidence in localization is increased when these two measures agree, and further testing is required when they do not.

Recording with *stereotactically implanted depth electrodes* (SEEG) is typically used in mesial temporal structures and inaccessible mesial and ventral cortical regions. Depth electrodes are in direct contact with neuronal generators of EEG signals, while the signals generated by a neuronal population relatively close to the electrode are undetectable. Therefore, it may be expected that many parts of the brain, even in relatively close proximity to electrode contacts, will remain unexplored. An important question is whether the electrographic ictal onset recorded with depth electrodes and accompanied by detectable epileptic clinical manifestations, is the real beginning of ictal events. In 1983,

Elger and Speckmann[12] reported that recordings performed during the interictal state of a developing experimental focus showed that in the period 400–500 ms prior to the epicortical seizure potential, a continuous increase of the mean neuronal discharge frequency develops until the PDS is initiated!

Depth electrode implantation requires that prior noninvasive evaluation reduce potential sites for an epileptogenic lesion to only a few, because the number of electrodes that can be inserted is limited. Stereotactic insertion of electrodes through twist drill holes is associated with a small but real risk of intracranial bleeding[42], infection and neurological deficits[43]. The significance of these potential risks is emphasized by the fact that the proportion of patients undergoing diagnostic chronic intracranial recordings who do not subsequently come to a therapeutic surgical procedure in some series is close to 50%[81] and on average is 25%[23, 42]. At present it is widely accepted[23] that the previous misconception should be dispelled that only intracranial EEG recordings provide reliable and definitive localizing evidence and that therefore in the presurgical evaluation of epileptic patients, they should probably always be used[65, 71].

Development in the techniques of *chronically implanted epidural and subdural strip, tress or grid electrodes* have not yet allowed sufficient data to accumulate. It seems that these techniques might provide better recordings from lateral neocortex than can be obtained with SEEG, but identification of epileptiform abnormalities from mesial, particularly limbic structures is often inadequate[36, 42].

Acute electrocorticography (ECoG) has been widely exploited at surgery in order to identify the location and extent of the epileptogenic brain tissue on the basis of EEG spike activity[45]. Since interictal spikes can be recorded widely, over large areas of brain, there is no clear indication of when these spikes identify cortex that must be removed, and when "spiking" cortex can be spared without affecting surgical outcome. However, this technique remains indispensable for delineation of a potentially epileptogenic cortical area, particularly in planning extratemporal cortical excisions[52].

e) *Brain mapping* is designed to identify functionally important brain areas that must be spared at resective surgery for epilepsy. The functions that are most often at risk are those related to sensorimotor areas, language areas in frontal and temporal portions of the dominant hemisphere, and areas important to memory in the temporal lobe. There are three tests that have been used as markers of cortical function: recording of spontaneous EEG activity[45], recording of cortical evoked potentials[35], and electrical stimulation[45]. The example of spontaneous activity with a constant anatomical distribution is the Mu rhythm, which usually is strictly localized to the motor hand area[36]. Unfortunately, this physiological rhythm is frequently not sufficiently well defined to be used as a reliable marker of cortical function. Of the different types of evoked potentials, only the somatosensory provides useful information when trying to define cortical function. They are reliable indices of the location of the rolandic fissure at the hand and foot area[25, 35]. They do not, however, define the anteroposterior extent of the sensorimotor area. The most powerful technique for localizing cortical function is direct electrical stimulation of the cortex. However, one should bear in mind that the effect of electrical stimulation is highly nonphysiological. The effect of a stimulus differs depending on whether a primary afferent/efferent area or an "association" area is stimulated. In primary afferent/efferent areas the main effect is "positive", producing muscle twitches in the motor area, parasthesiae in the sensory area, and simple visual or auditory hallucinations in the corresponding receiving areas. There is, however, evidence that positive symptoms are usually superimposed on the inhibitory effects: stimulation of the primary auditory cortex produces deafness, of the calcarine fissure produces blindness[44], and of the primary somatosensory area produces hypaesthesia[34].

The stimulation effect in "association" area is "negative". It becomes obvious when the patient is asked to, but is unable to perform, a specific task that requires the participation of the cortex that is being stimulated. The typical example is inability to talk when the language area is stimulated. On the basis of surgical experience Rasmussen pointed out[51] that failure to elicit speech interference with cortical stimulation does not, unfortunately, guarantee that the cortex stimulated is not involved in speech function. Rasmussen advises that even if no aphasic response is obtained with stimulation of the frontal opercular gyri in a hemisphere known to be dominant for speech, at least the first three opercular gyri in front of the lower end of the precentral gyri must be carefully preserved. On the other hand, arrest of speech may be obtained when stimulating cortex more superiorly on the lateral hemisphere[78] in the area proved to be dispensable. Nevertheless, mapping of cortical speech representation by electrical stimulation in temporal and parietal regions with the patient under local anesthesia, seems to be a reliable method. Recently, some authors suggested that there was a lateral cortical component to recent verbal memory (in addition to the traditional components in medial temporal lobe and around the third ventricle) and that the portions of lateral temporal lobe related to memory could be identified by stimulation mapping[41].

III. Is the Present C-O Discrepancy Justifiable?

Surgical outcome with respect to epileptic seizures, side effects, and quality of life is the best indicator of the justification for epilepsy surgery. Engel recently reported a survey of data on *surgical outcome* with respect to epileptic seizures gathered from forty known active epilepsy surgery centres[14] (Table 1).

This survey included only patients with follow up of more than 2 years. In the group of 2,336 anterior temporal lobectomies, 55.5% were seizure free, 27.7% were improved, and 16.8% were not improved. In the group of 825 extratemporal resections, 43.2% were seizure free, 27.8% improved, and 29,1% not improved. In the group of 88 hemispherectomies, 77.3% were seizure free, 18,2% improved and 4.5% were not improved.

Table 1. *Surgical Outcome with Respect to Epileptic Seizures from 40 Active Epilepsy Centres* (follow-up more than 2 years) (Engel, J., Jr., 1987)

Operative procedure	No.	Seizure free	Improved	Not improved
Anterior temporal lobectomies	2336	55.5%	27.7%	16.8%
Extratemporal resections	825	43.2%	27.8%	29.1%
Hemispherectomies	88	77.3%	18.2%	4.5%

IV. Results of Personal Series

My personal series was not included in the cited survey and will be reported here.

Our preoperative protocol included: case history, neurology, neuroophthalmology, neuropsychology, in selected cases Amytal-test for speech lateralization, CT scan, and occasionally angiography and electrographic recordings. Our protocol of EEG examination included: scalp electrodes, sphenoidal electrodes, orbital electrodes[59], chronically implanted subdural electrodes (only in cases presumed to have an epileptogenic focus in the vicinity of sufficiently functioning sensorimotor

cortex), electrocorticography and electrical stimulation for brain mapping (with subdural electrodes and/or at operation under local and neurolept analgesia). In the majority of operated patients we recorded only interictal epileptic abnormalities (in awake and pharmacologically induced sleep, recording repeated 3 or more times with many montages). In rare selected patients we used long term EEG recording with the holter technique. In the CPS group our indication for operation in the great majority of cases were the presence of unilateral epileptic spikes or, occasionally, bilateral independent spikes that were predominantly unilateral, accompanied by some other congruent lateralizing finding: unilateral morphological lesion visible on CT scan, neuropsychological finding, or lateralizing clinical manifestation.

I have operated on 300 patients for drug resistant epileptic fits since 1974. The great majority of these patients were operated on after my training at the Montreal Neurological Institute in 1980.

In the last two years 246 selected candidates were admitted for epilepsy surgery, and after the presurgical evaluation we operated on 105 patients (43%).

In the series of 300 patients operated on, 224 (75%) had CPS and 76 (25%) belonged to some other epileptic group. In this non-temporal group there were 53 (70%), patients with extratemporal focus, 4 (5%) with two or three separated unilateral foci, 16 (21%) with infantile hemiplegia and epilepsy, and 3 (4%) with epilepsia partialis continua-Kojewnikow.

In the CPS-epilepsy group we have three standardized surgical approaches, which depend on the side and "type" of the electrical field distribution of interictal epileptic abnormalities recorded with sphenoidal and orbital electrodes. In the cases of the "temporobasal" type of interictal abnormalities (maximum of electrical field recorded with sphenoidal electrode) we performed temporal lobectomy (TL) (on the nondominant side 4.5 cm, on the dominant side 3 cm from the temporal pole, measured along the superior temporal margin) with amygdalectomy, uncusectomy and resection of pes hippocampi-1 cm). In the cases of "diffuse" type of interictal abnormalities (temporal baso-convexity field equipotentiality) on the nondominant side we perform ECoG for determining the posterior extent of the convexity resection, on the dominant side resection is similar to "temporobasal" type. In the cases of "frontobasal" type of interictal abnormalities (maximum of electrical field recorded with orbital electrode[59]) we perform on both dominant and nondominant side a "Restricted Temporo-Frontal Resection" (RTFR)

Table 2. *Surgical Outcome in Patients with CPS*

Operation	No. of patients	Seizure free	Improved	Not improved
All "types" (follow-up 2–15 years)	126 (100%)	90 (71%)	21 (17%)	15 (12%)
R.T.F.R. (follow-up 1–6 years)	51 (100%)	45 (88%)	4 (8%)	2 (4%)

Table 3. *Surgical Outcome in Patients with Extratemporal Epileptogenic Focus* (follow-up 2–12 years)

No. of patients	Seizure free	Improved	Not improved
25 (100%)	14 (56%)	9 (36%)	2 (8%)

(temporal polectomy, 2.5 to 3 cm from the temporal pole, amygdalectomy, resection of pes hippocampi, uncusectomy, and resection of posterior 2/3 of frontoorbital cortex[59]). In the last six years (since we use orbital electrodes) we have operated on 172 patients for CPS. Temporal lobectomy (TL) was performed in 98 (57%) and RTFR was performed in 74 (43%) patients.

Taking into account all patients operated on for CPS, there are 126 patients with a follow up of 2 to 15 years (mean 4.5 years) (Table 2).

In this series 90 (71%) are seizure free, 21 (17%) are improved, and 15 (12%) are not improved. Much better results are seen with the cases of RTFR: there are 51 patients with followup 1 to 6 years (mean 3.5 years), 45 (88%) are seizure free, 4 (8%) are improved, and 2 (4%) are not improved.

In the group of 25 patients with an extratemporal epileptogenic focus (Table 3) with follow up of 2 to 12 years (mean 5.3 years), 14 (56%) are seizure free, 9 (36%) are improved and 2 (8%) are not improved.

In the group of 11 patients with infantile hemiplegia and epilepsy (Table 4) and follow up of 2 to 13 years (mean 6 years), submitted to partial hemispherectomy, 7 (64%) are seizure free, 2 (18%) are improved and 2 (18%) are not improved.

In the group of 4 patients with two or three unilateral independent potentially epileptogenic foci with follow up of 7 to 10 years (Table 4), 2 (50%) are seizure free and are improved.

In the group of 3 patients with epilepsia partialis continua-Kojewnikow (Table 4), with follow up of 1 to 2 years, 2 are seizure free, and 1 is patient.

The important aspect of any surgical procedure is its *potential risks and side effects*. Recently, Van Buren reported a survey of the literature on the complications of surgical procedures in the diagnosis and treatment of epilepsy[77] (Table 5). In the series of 1,911 anterior temporal lobectomies there was less than 0.5% mortality and 5% morbidity; in the series of 432 extratemporal resections there was 0% mortality and 6% morbidity; in the series of 84 hemispherectomies, there was 4% mortality and 17% morbidity; in the series of 130 corpus callosum sections, there was 2% mortality and 11% morbidity.

Table 5. *Complication of Surgical Procedures Survey of the Literature* (van Buren, J. M., 1987)

Operation	No. of patients	Mortality	Morbidity
Temporal lobectomies	1911	< 0.5%	5%
Extratemporal resections	432	0%	6%
Hemispherectomies	84	4%	17%
Corpus callosum sections	130	2%	11%

Table 4. *Surgical Outcome in Specific Epileptic Groups*

Group	Follow-up	No. of patients	Seizure free	Improved	Not improved
Infantile Hemiplegia	2–13 years	11 (100%)	7 (64%)	2 (18%)	2 (18%)
2 or 3 Unilateral foci	7–10 years	4	2	2	0
EPI Kojewnikow	1–2 years	3	2	1	0

In my series of 300 patients operated on for drug resistant epilepsy, one retarded and hemiplegic 5-year-old boy passed away suddenly, for an unknown reason, on the first day following the partial hemispherectomy. There was no any other major permanent complication.

The main concern regarding the side effects of temporal lobectomy has been concentrated on *recent memory deficits*. It has been suggested that this memory problem is so pervasive after anatomically standard temporal lobectomy in the dominant hemisphere that such operations should not be done in anyone whose occupation is dependent on a facile memory[7]. The hippocampus is the temporal lobe structure usually considered to be essential for memory, with intact memory requiring one intact hippocampus[46]. The usual approach to avoid memory deficits after temporal lobectomy is to assess the effect of intracarotid Amytal perfusion on the side of the proposed resection on a measure of recent memory[38]. However, postoperative recent memory deficits have been reported in patients managed in this manner, with a mean reduction in the Wechsler Verbal Memory score 1 year after left temporal resection in seizure free patients of 18% in one series[57] and 30% in another[40]. Recently, Ojeman and Dodrill[82] demonstrated that there was a lateral cortical component to recent memory. They suggested that damage to that lateral cortical component accounted for some of the memory deficit after temporal lobectomy, and that the portions of lateral temporal lobe related to memory could be identified by stimulation mapping. Whether memory deficits are absent if hippocampus is resected but the lateral temporal cortical memory component preserved is presently unknown. Memory deficits are reported to be minimal in patients undergoing amygdalohippocampectomy[82]. However, one case of global amnesia has been reported following amygdalohippocampectomy[56].

V. Final Comments

The outcome of epilepsy-surgery with respect to epileptic seizures, side effects and quality of life is very satisfactory. In spite of many missing pieces in the epilepsy puzzle and defects in diagnostic procedures, there is now such an extensive experience with epilepsy-surgery that we can safely say that, under the appropriate circumstances, it is as routine as any other classical neurosurgical procedure. It is not humane to deprive a great number of suffering epileptic patients of the chance to transform their desparate "hell of a life"

into a "promising paradise" of psychosocial adaptation.

The epilepsy service in each country should be organized in the hierarchy mode, with an agreed doctrine with respect to drug treatment, criteria for the identification of the candidates for presurgical evaluation and the principles of psychosocial rehabilitation. The basic and most important level in an epilepsy-service comprises neurologists, psychologists and psychiatrists, who treat and follow up epileptic patients for years and identify the candidates for presurgical evaluation. The higher level of an epilepsy-service comprises epilepsy-centres that concentrate on the equipment and multidisciplinary experts, as well as neurosurgical centres capable of performing multidisciplinary presurgical evaluation, surgical interventions, and research work. Such research should lead to the understanding of the brain, the devising of new diagnostic and surgical procedures and defining the specific indications for particular tests and surgical procedures.

The success of epilepsy-surgery very much depends on the quality of the neurosurgeon's participation in all stages of the presurgical and surgical procedures. He is the axle and coordinator of the team-work in preoperative evaluation, as well as the executor of the well thought-out therapeutic "brain cutting". To meet these demands, the neurosurgeon must be well educated and trained in "classical" neurosurgery as well as in the specific diagnostic and surgical procedures relevant to epilepsy surgery.

References

1. Armstrong DD, Bruton CJ (1987) Postscript: what terminology is appropriate for tissue pathology? How does it predict outcome? In: Engel J Jr (ed) Surgical treatment of the epilepsies. Raven Press, New York, pp 541–552
2. Berger H (1979) Über das Elektroenzephalogramm des Menschen. Arch Psychiat Nerv Krankh 87: 527
3. Broca P (1878) Anatomie comparée circonvolutions cérébrales. Le grand lobe limbique et la scissure limbique dans la série des manifères. Rev anthropol Ser 1: 384–498
4. Crandell PH (1977) Role of neurosurgery in management of medication-resistant epilepsy. In: Plan for nationwide action on epilepsy, Vol 2. DHEW Publications, Washington, DC, pp 327–334
5. Dandy WE (1918) Ventriculography following injection of air into the cerebral ventricles. Ann Surg 66. 5
6. Dandy WE (1919) Roentgenography of the brain after the injection of air into the spinal canal. Ann Surg 70: 397
7. Delgado-Escuenta A, Walsh G (1983) The selection process for surgery of intracable complex partial seizures: Surface EEG and depth electrography. In: Ward A, Penry J, Purpura D (eds) Epilepsy. Raven Press, New York

8. Delgado-Escuenta VA, Ward AA Jr, Woodbury DM, Porter RJ (1986) New wave of research on the epilepsies. In: Delgado-Escuenta VA, Ward AA Jr, Woodbury DM, Porter RJ (eds) Advances in neurology, Vol 44. Raven Press, New York, pp 1–55

9. Dreifuss EF (1987) Goals of surgery for epilepsy. In: Engel J Jr (ed) Surgical treatment of the epilepsies. Raven Press, New York, pp 31–49

10. Dreifuss EF, Martinez-Lage M, Roger J, Seino M, Wolf P, Dam M (1985) Proposal for classification of epilepsies and epileptic syndromes. Epilepsia 26: 268–278

11. Elger EC (1987) Basic epileptology. In: Wieser GH, Elger CE (eds) Presurgical evaluation of epileptics. Springer, Berlin Heidelberg New York, pp 29–34

12. Elger EC, Speckmann EJ (1983) Penicillin induced epileptic foci in the motor cortex: vertical inhibition. Electroencephalogr Clin Neurophysiol 56: 604–622

13. Engel J Jr (1987) Preface. In: Engel J Jr (ed) Surgical treatment of epilepsies. Raven Press, New York

14. Engel J Jr (1987) Outcome with respect to epileptic seizures. In: Engel J Jr (ed) Surgical treatment of epilepsies. Raven Press, New York, pp 553–571

15. Engel J Jr, Cahan LD, Sutherling WW, Crandall PH, Phelps ME (1987) The use of positron emission tomography in the surgical treatment of epilepsy. In: Wieser GH, Elger CE (eds) Presurgical evaluation of epileptics. Springer, Berlin Heidelberg New York, pp 136–139

16. Engel J Jr, Rausch R, Lieb PJ, Kuhl ED, Vrandall HP (1981) Correlation of criteria used for localizing epileptic foci in patients considered for surgical therapy of epilepsy. Ann Neurol 9: 215–224

17. Foerster O (1925) Zur Pathogenese und chirurgischen Behandlung der Epilepsie. Zentralbl Chir 52: 531–549

18. Foerster O, Altenburger H (1935) Elektrobiologische Vorgänge an der menschlichen Hirnrinde. Dtsch Z Nervenheilk 135: 277–288

19. Foerster O, Penfield W (1930) The structural basis of traumatic epilepsy and result of radical operation. Brain 53: 99–120

20. Fritsch G, Hitzig E (1870) Über die elektrische Erregbarkeit des Grosshirns. Arch Anat Physiol 37: 300

21. Gastaut H (1953) So-called "psychomotor" and "temporal" epilepsy. Epilepsia 5: 59–99

22. Gastaut H (1970) Clinical and electroencephalographic classification of epileptic seizures. Epilepsia 11: 102–113

23. Gloor P (1987) Commentary: Approaches to localization of the epileptogenic lesion. In: Engel J Jr (ed) Surgical treatment of the epilepsies. Raven Press, New York, pp 97–100

24. Gloor P (1987) Volume conductor principles: Their application to the surface and depth electroencephalogram. In: Wieser GH, Elger DE (eds) Presurgical evaluation of epileptics. Springer, Berlin Heidelberg New York, pp 59–68

25. Goldring S, Gregorie ME (1984) Surgical management of epilepsy using epidural recordings to localize the seizure focus. J Neurosurg 60: 457–466

26. Goldensohn ES, Purpura PD (1963) Intracellular potentials of cortical neurons during focal epileptogenic discharges. Science 139: 840–842

27. Hauser WA, Kurland TL (1975) The epidemiology of epilepsy in Rochester, Minnesota 1935 through 1967. Epilepsia 16: 1–66

28. Horsley Sir V (1886) Brain-surgery. Br Med J 2: 670–675

29. Hounsfield GN (1973) Computerized transverse axial scanning (tomography). Part I. Description of system. Br J Radiol 46: 1016–1022

30. James EA Jr, Partain C (eds) (1981) Proceedings: nuclear magnetic resonance (NMR) imaging symposium, Vanderbilt University, Nashville, TN, October 26–27, 1980. J Comput Assist Tomogr 5: 285–305

31. Jasper HH (1949) Etude anatomo-physiologique des épilepsies. 2e Congrès Intern. d'EEG, Paris. Electroencephalogr Clin Neurophysiol [Suppl] 2

32. Jasper HH (1949) Electrocorticograpm in man. Electroencephalogr Clin Neurophysiol [Suppl] 2: 16–29

33. Lanterbur PC (1973) Image formation by induced local interactions: examples employing NMR. Nature 242: 190–191

34. Libet B, Alberts WW, Wright EW, Feinstein B (1972) Cortical and thalamic activation in conscious sensory experience. In: Sonjen GG (ed) Neurophysiology studied in man. Excerpta Medica, Amsterdam, pp 157–168

35. Lüders H, Dinner DS, Lesser PL, Morris HH (1986) Evoked potentials in cortical localization. J Clin Neurophysiol 3: 75–84

36. Lüders H, Lesser RP, Dinner DS, Morris HH, Hahn JF, Friedman L, Shipper G, Wyllie E, Friedman D (1987) Commentary: chronic intracranial recording and stimulation with subdural electrodes. In: Engel J Jr (ed) Surgical treatment of the epilepsies. Raven Press, New York, pp 297–321

37. Matsumoto H, Ajmone-Marsan C (1964) Cortical cellular phenomena in experimental epilepsy: interictal manifestations. Exp Neurol 9: 305–326

38. Milner B, Branch C, Rasmussen T (1962) Study of short-term memory after intracarotid injection of sodium amytal. Trans Am Neurol Assoc 87: 224–226

39. Moniz E (1927) L'encephalographie artériell som importance dans la localization des tumeurs cérébrales. Rev Neur (Paris) 2: 72–90

40. Novelly R, Augustine E, Mattson R et al (1984) Selective memory improvement and impairment in temporal lobectomy for epilepsy. Ann Neurol 15: 64–67

41. Ojemann G, Dodrill C (1985) Verbal memory deficits after left temporal lobectomy for epilepsy: Mechanism and intraoperative prediction. J Neurosurg 62: 101–107

42. Ojeman AG, Engel J Jr (1987) Acute and chronic intracranial recording and stimulation. In: Engel J Jr (ed) Surgical treatment of epilepsies. Raven Press, New York, pp 263–288

43. Olivier A, Gloor P, Quesney F, Anderman F (1983) The indications for and the role of depth electrode recording in epilepsy. Appl Neurophysiol 46: 33–36

44. Penfield W (1958) The excitable cortex in conscious man. Liverpool University Press, Liverpool

45. Penfield W, Jasper HH (1954) Epilepsy and the functional anatomy of the human brain. Little, Brown, Boston

46. Penfield W, Mathieson G (1974) Memory. Autopsy findings and comments on the role of hippocampus in experiential recall. Arch Neurol 31: 145–154

47. Penfield W, Rasmussen T (1957) The cerebral cortex of man. Macmillan, New York

48. Perot P, Weir B, Rasmussen T (1966) Tuberous sclerosis: Surgical therapy for seizures. Arch Neurol 15: 498–506

49. Porter RJ, Wolf AA, Penry KJ (1971) Human electroencephalographic telemetry. Am J EEG Technol 11: 145–159

50. Rasmussen T (1975) Cortical resection in the treatment of local epilepsy. In: Pupura JK, Penry JK, Walter RD (eds) Neurosurgical management of the epilepsies. Advances in neurology, vol 8. Raven Press, New York, pp 139–154

51. Rasmussen T (1975) Surgery of frontal lobe epilepsy. In: Purpura JK, Penry JK, Walter RD (eds) Neurosurgical management of the epilepsies. Advances in neurology, vol 8. Raven Press, New York, pp 197–205

52. Rasmussen T (1987) Commentary: extratemporal cortical excisions and hemispherectomy. In: Engel J Jr (ed) Surgical treatment of the epilepsies. Raven Press, New York, pp 417–424

53. Rasmussen T, Mc Cann W (1968) Clinical studies of patients with focal epilepsy due to "chronic encephalitis". Trans Am Neurol Assoc 93: 89–94

54. Rausch R (1985) Differences in memory function with right and left temporal lobe dysfunction. In: Benson FD, Zaidel E (eds) The dual brain. Guilford Press, New York, pp 247–261

55. Rausch R (1987) Psychological evaluation. In: Engel J Jr (ed) Surgical treatment of the epilepsis. Raven Press, New York, pp 181–195

56. Rausch R, Babb LT, Brown JW (1986) A case of amnestic syndrome following selective amygdalohippocampectomy. Presented at the International Neuropsychological Society Meeting, Danver, February

57. Rausch R, Crandall P (1982) Psychological status related to surgical control of temporal lobe seizures. Epilepsia 23: 191–202

58. Rayport M (1977) Role of neurosurgery in management of medication-resistant epilepsy. In: Plan for Nationwide Action on Epilepsy, vol 2. DHEW Publications, Washington DC, pp 314–324

59. Ribarić I, Sekulović N (1989) Experience with orbital electrodes in the epilepsy surgery-results of temporofrontal resections. Acta Neurochir (Wien) [Suppl 46]: 21–24

60. Robitaille Y (1987) Pathological changes relevant for seizure generation. In: Weiser GH, Elger CE (eds) Presurgical evaluation of epileptics. Springer, Berlin Heidelberg New York, pp 79–85

61. Rodin EA (1968) The prognosis of patients with epilepsy. C Thomas, Springfield, Ill

62. Rovit RR, Gloor P, Henderson L (1960) Temporal lobe epilepsy study using multiple basal electrodes. I. Description of method. Neurochirurgia 3: 6–19

63. Sano K, Malamud N (1953) Clinical significance of sclerosis of the cornu amonis. Arch Neurol Psychiatr 70: 40–53

64. Schaltenbrand G, Bailey P (eds) Introduction to stereotaxis with an atlas of the human brain, Vol I-III. Georg Thieme, Stuttgart

65. Spencer SS (1981) Depth electroencephalography in selection of refractory epilepsy for surgery. Ann Neurol 9: 207–214

66. Sperling MR, Mendius RJ, Engel J Jr (1986) Mesial temporal spikes: a simultaneous comparison of sphenoidal, nasopharyngeal, and ear electrodes. Epilepsia 27: 81–86

67. Sperling MR, Sutherling WW, Nuwer MR (1987) New techniques for evaluating patients for epilepsy surgery. In: Engel J Jr (ed) Surgical treatment of the epilepsies. Raven Press, New York, pp 235–257

68. Sperling MR, Wilson G, Engel J Jr, Babb LT, Phelps M, Bradley W (1986) Magnetic resonance imaging in intractable partial epilepsy: Correlative studies. Ann Neurol 20: 57–62

69. Stevens JR (1969) Localization of epileptic focus by protracted monitoring of EEG by radio telemetry. Epilepsia 10: 420

70. Storm van Leeuwen W, Kamp A (1969) Radio telemetry of EEG and other biological variables in man and dog. Proc Royal Soc Med 62: 451–453

71. Talairach J, Bancaud J, Szikla G, Bonis A, Geier S, Vadreune C (1974) Approche nouvelle de la neurochirurgie de l'epilepsie. Méthodologie stéréotaxique et resultats thérapeutiques. Neurochirurgie 20 [Suppl 1]: 240

72. Talairach J, David M, Touruoux P, Corredor H, Kvasina T (1957) Atlas d'anatomie stéréotaxique des noyaux gris centraux. Masson, Paris

73. Talairach J, Szikla G, Touruoux P et al (1967) Atlas d'anatomie stéréotacique du téléncephale. Masson, Paris

74. Taylor J (ed) (1932) Selected writings of John Hughlings Jackson, Vol 1. On Epilepsy and epileptiforme convulsions. Hodder and Stoughton, London

75. Taylor DC, Falconer AM, Bruton JC, Correllis AN (1971) Focal dysplasia of cerebral cortex in epilepsy. J Neurol Neurosurg Psychiatry 34: 369–387

76. Theodore WH, Brooks R, Sato S, Patronas N, Margolin R, Di Chiro G, Porter JR (1984) The role of positron emission tomography in the evaluation of seizure disorders. Ann Neurol 15 [Suppl]: S 176–S 179

77. Van Buren MJ (1987) Complications of surgical procedures in the diagnosis and treatment of epilepsy. In: Engel J Jr (ed) Surgical treatment of the epilepsies. Raven Press, New York, pp 465–475

78. Van Buren MJ, Fedio D (1976) Functional representation on the medial aspect of the frontal lobes in man. J Neurosurg 44: 275–289

79. Walker AE, Marshal C, Beresford NE (1947) Electrocorticographic characteristics of the cerebrum in posttraumatic epilepsy. Res Publ Assoc Res Nerv Ment Dis 26: 502–515

80. Wieser HG (1987) Stereo-electroencephalography. In: Wieser GH, Elger CE (eds) Presurgical evaluation of epileptics. Springer, Berlin Heidelberg New York, pp 192–204

81. Wieser HG, Bancaud J, Talairach J, Bonis A, Szikla G (1979) Comparative value of spontaneous and chemically and electrically induced seizures in establishing the lateralization of temporal lobe seizures. Epilepsia 20: 47–60

82. Wieser HG, Yasargil M (1982) Selective amygdalohippocampectomy as a surgical treatment of mesiobasal limbic epilepsy. Surg Neurol 17: 445–457

83. Willis RA (1958) The borderland of embryology and pathology. Butterworth and Co Ltd, London

84. Williamson DP, Wieser HG, Delgado-Escuenta AV (1987) Clinical characteristics of partial seizures. In: Engel J Jr (ed) Surgical treatment of the epilepsies. Raven Press, New York, pp 101–120

85. Woodbury LA (1978) Incidence and prevalence of seizure disorders including the epilepsies in the United States of America. In: The commission for the control of epilepsies and its consequences, vol IV. DHEW Publication (NIH) Washington DC, pp 78–276

Correspondence: I. I. Ribarić, M.D. Neurosurgical Clinic, University Clinical Centre, Visegradska 26, 11000-Beograd, Yugoslavia.

Acta Neurochirurgica, Suppl. 50, 117–118 (1990)

Why Are so Few Patients with Epilepsy Treated Surgically?

A United Kingdom Perspective

D. Chadwick

Regional Neurological Center, Walton Hospital, Liverpool, U. K.

Summary

The reasons why many patients with drug resistant complex partial epilepsy who might benefit from surgery are not appropriately assessed are examined, and include the attitudes and prejudices of Neurologists, Neurophysiologists, Neuroradiologists, Neurosurgeons and the patients themselves. Finally, the implications of provision of the necessary resources in the United Kingdom are presented.

Keywords: Epilepsy; operative treatment.

There can be no doubt that many patients, particularly those with drug resistant complex partial epilepsy who would be expected to benefit from anterior temporal lobectomy or amygdalo-hippocampectomy, are never assessed for surgical treatment let alone offered it. With estimates that between one and two patients per 100,000 of the general population developing epilepsy would benefit from such surgery (see pp 32–37) and with the very acceptable results of surgery in otherwise very disabled patients, it is important to consider why so few patients are operated on and how this situation might be improved.

The following discussion represents a personal view based on knowledge of the services to people with epilepsy in the United Kingdom though many of the points may to a greater or lesser degree be applicable to other European countries. It would seem that problems exist in a number of key areas:

Neurological Attitudes

The situation in the United Kingdom is somewhat unusual in that many adolescent and adult patients with epilepsy may be cared for by general physicians rather than specialist neurologists. This group of physicians is largely unaware of the potential benefits of surgery in epilepsy. However, it is still relatively rare for neurologists caring for patients with epilepsy to refer them for full appraisal for surgery, with the exception of those patients in whom routine neuro-imaging procedures disclose the obvious presence of a cerebral tumour. This probably reflects the fact that relatively few British neurologists have a specific interest or would claim a particular expertise in epilepsy. Specialist epilepsy services are rare in the United Kingdom and patients attending neurological clinics may frequently be seen by junior medical staff training in neurology. These are again likely to be a group of clinicians who cannot be expected to be effective in selecting patients for surgical treatment.

There is no doubt that an increase in surgical treatment will be largely dependent on increasing the proportion of patients with epilepsy who are seen and managed by neurologists with a specific interest in epilepsy, ideally within the setting of a multidisciplinary epilepsy clinic.

Neurophysiological Attitudes

In the United Kingdom as neurophysiology has become established as an independent neuroscience subspeciality, this has coincided with the development of interest in the peripheral nervous system and studies of nerve conduction and electromyography. The study and development of techniques for evoked potential monitoring has also tended to detract from an interest in EEG services for patients with epilepsy. Thus, only a small proportion of neurophysiologists active in the United Kingdom have a major interest in epilepsy or significant clinical exposure to the care of patients with epilepsy. Overall, the provision of EEG services in the United Kingdom is extremely poor and many misunderstandings still exist over the use and value of the EEG in the diagnosis of epilepsy. Within this setting there is relatively little expertise available in the very highly specialised techniques of long-term monitoring and depth and other intracranial recordings necessary to support surgical epilepsy programmes.

It is clear that the subspeciality of neurophysiology will need major expansion and more extensive training programmes in order to fulfil the goals of expanding the availability of surgical treatment of epilepsy.

Neuroradiology

The important contributions of neuroradiology scarcely need to be emphasized and the potential for identifying patients suitable for treatment is clearly demonstrated by Anslow (see pp 76–79). Whilst most neuroradiological centres will successfully detect major tumours by routine CT scanning, many more subtle lesions can be demonstrated by more directed and sensitive CT studies. The provision of magnetic resonance does seem a major benefit in the assessment of patients, paticularly as it may identify structural lesions missed on CT scanning and also delineate atrophic and gliotic pathology more acurately than CT. As this technology becomes more generally available, its presence may well have a significant impact on increasing the potential for surgical treatment of epilepsy without exhaustive and intricate neurophysiological study. The role of functional imaging techniques awaits definition. SPECT is a relatively cheap technology and may become generally available for the evaluation of patients with epilepsy. Its precise role has yet to be defined. In contrast, PET imaging is expensive and is unlikely to be available to other than to highly specialist centres. The potential of magnetic resonance spectroscopy is yet to be explored but as the technology develops it is possible that this could also become more generally available and facilitate the selection of more patients for treatment by surgery.

Neurosurgical Attitudes

There are only a few young neurosurgeons in training who have an interest in epilepsy surgery. Traditionally neurosurgeons see only a highly selected population of patients with epilepsy, largely those with cerebral tumours associated with epilepsy. Functional neurosurgery seems only to be in its infancy and there is some reluctance on the part of many neurosurgeons to consider resection of cerebral tissue that may not harbour macroscopically obvious pathology. It is clear that more neurosurgical centres must develop a major interest in epilepsy if the potential demand for this treatment is to be satisfied. Neurosurgeons with expertise in the treatment of epilepsy need to extend and offer advice to other neurosurgical centres. As the success of surgical treatment is dependent on experience and a regular thoughput of patients, the situation in which an individual surgeon may only perform one or two operations for epilepsy each year is to be avoided.

Patients' Attitudes

Those concerned with the care of epilepsy realise that there is a considerable part to be played by the physician in the education of patients with epilepsy. The epileptic patient as a consumer of medical services clearly needs to be more aware that surgical treatment can be highly successful in carefully selected groups of patients. Should patients and patient organisations appreciate this more widely then greater pressure will be brought to bear on clinicians and health care agencies to promote the development of services.

Resource Implications

It would not be unreasonable to suggest that on the basis of present figures the number of patients being offered surgery in most European countries should increase two or three fold during the next decade. As the adequate assessment for surgery demands a multidisciplinary team including neurologists, neurophysiologists, neuroradiologists, neuropsychologists and neurosurgeons, epilepsy surgery programmes clearly need to be based in regional units in which such services are centred. The organisation of regional and suprarregional services within individual countries needs to be addressed. It may not be necessary for all centres offering a surgical programme to patients with epilepsy to have facilities to provide every conceivable investigation up to the point of implantation of electrodes for intracranial recording, but it would seem reasonable that each European country has at least one centre with such a level of expertise which routinely undertakes such studies. The projected costs of the provision of the human and capital resources necessary to expand epilepsy surgery have major implications for health care system in individual countries. However, the cost effectiveness of epilepsy programmes cannot be in doubt (pp 100–106). Pressure must now be exerted at a national and European level to make the necessary resources available.

Correspondence: D. Chadwick, DM, Regional Neurological Center, Walton Hospital, Rice Lane, Liverpool L9 1AE, U.K.

Acta Neurochirurgica, Suppl. 50, 119–121 (1990)

Temporal Lobectomy

J. Bidzinski

Department of Neurosurgery, Warsaw Medical Academy, Warsaw, Poland

Summary

The indications for, and results of, the author's personal series of 486 resective operations in epileptic patients are reviewed, together with the long term results of 286 patients undergoing temporal lobectomy. Almost 50% of such patients were seizure free when followed for two to thirty years.

Keywords: Epilepsy; temporal lobe; operative treatment; results.

The surgical treatment of epilepsy started in Warsaw in 1957 and we have now performed 486 resective operations in epileptic patients up to the end of 1988 (Table 1). Temporal lobe epilepsy was present in 66% of the surgically treated patients. In more than 80% of the group of 88 patients with gross epileptogenic lesions, the temporal lobe was involved and resected, but due to electro-corticogram changes, the resection had to be extended to another lobe (whole or part). Hence, this group of patients is not included in this presentation.

Indications for Temporal Lobectomy

The indications for surgical intervention were:
1. Frequent epileptic fits
2. Resistance to pharmacological treatment
3. Localized temporal epileptogenic focus
4. Rapid mental deterioration.

The protocol for the pre-operative investigation of these patients for temporal lobectomy is outlined in Table 2.

The patients were observed in the Department of Neurosurgery on several occasions. In recent years, compliance with the prescribed drug treatment was monitored for at least six months to be sure that such drug therapy was ineffective.

Patient Population

The youngest patient was one year of age and the oldest was forty-seven years old (Table 3). The male to female ratio was about 1:1.

Table 2. *Pre-operative Investigations for Temporal Lobe Epilepsy*

1. Case history
2. Seizure pattern
3. Serial EEG recordings
4. Plain X-ray
5. Pneumoencephalogram or CT scan
6. Angiography (not routine)
7. Amytal test (routine)
8. Psychological assessment

Table 1. *Surgical Treatment of Epilepsy 1957–1988*

Resective surgery	
Temporal	320
Frontal	41
Parietal	10
Occipital	12
Rolandic	15
Gross epileptogenic lesion/more than one lobe	88
Total	486

Table 3. *Age of the Patients with Temporal Lobe Epilepsy*

Age	Number
0–5	4
6–10	11
11–15	36
16–20	84
21–25	74
26–30	54
31–35	25
36–40	19
41–	13

Table 4. *Temporal Lobe Epilepsy – Age of Onset*

Age	Number
0–5	105
6–10	62
11–15	73
16–20	41
21–25	24
26–	15

Table 5. *Duration of Temporal Lobe Epilepsy*

Years	Number
0–5	56
6–10	92
11–15	76
16–20	59
21–	37

Table 6. *Temporal Lobe Epilepsy – Probable Aetiology Based on the Case History*

Birth trauma	54
Head trauma	57
Infection	58
Not known	151

Table 7. *Temporal Lobectomy: Results in 286 Patients Followed for 2–30 Years*

	Number of patients	%	%
Very good	140	49.0	67.5
Good	53	18.5	
Poor	35	12.2	32.5
Bad	58	20.3	
	286		

In most cases, epilepsy started in early childhood (Table 4) and the duration of epilepsy was usually over five years (Table 5). This long period of conservative management reflected a limited understanding of the possible benefits of the surgical treatment of epilepsy among neurologists and patients. The probable aetiology of the temporal lobe epilepsy as extracted from the case history (Table 6) was doubtful in many cases, and, in older patients, there was no available data about birth and early childhood.

Results

In all patients, a standard temporal lobectomy was performed, including excision of the hippocampus and

Table 8. *Temporal Lobectomy: Results for Different Patterns of Seizure*

Seizure pattern	Results	
	Satisfactory	Unsatisfactory
Temporal with or without secondary generalization	84%	16%
Temporal and independent other	54%	46%

Table 9. *Temporal Lobectomy: Effect of Mental Retardation on Results*

I.Q.	Results	
	Satisfactory	Unsatisfactory
⩾ 70	82%	18%
⩽ 69	50%	50%

amygdala[1]. Three patients died (mortality 0.9%) and in two patients, a permanent hemiparesis resulted (morbidity 0.6%). Of the 318 patients discharged from hospital, 20 were lost to follow-up and follow-up was less than two years in 12 other patients, leaving 286 patients who had been followed from two to thirty years.

The following outcome scale was used:

1. Very good result – no seizures
2. Good result – two seizures or very infrequent non-incapacitating auras
3. Poor result – less frequent seizures and some improvement in social life
4. Bad result – no improvement.

In order to analyse the results statistically, outcomes 1 and 2 were regarded as satisfactory, and outcomes 3 and 4 unsatisfactory. The results in the 286 patients with long follow-up are presented in Table 7. In patients with temporal lobe seizures, the results were better than in patients with temporal and other independent types of seizure (Table 8). The results were much worse in patients with mental deterioration than in patients with an IQ over 70 (Table 9).

Conclusion

Temporal lobe epilepsy that is resistant to pharmacological treatment should be treated by temporal lobectomy. In 67% of such patients a satisfactory result can be obtained, and almost 50% of patients are seizure-free when followed for two to thirty years. As the results in mentally retarded patients are worse, psychological deterioration during drug treatment of temporal lobe epilepsy is another indication for surgery.

Reference

1. Bidzinski J (1987) Anterior Temporal Lobectomy at the Neurosurgical Clinic, Warsaw. In: Engel J Jr (ed) Surgical treatment of the epilepsies. Raven Press, New York, pp 647–651

Correspondence: Prof. J Bidzinski, Department of Neurosurgery, Warsaw Medical Academy, Banacha Str. 1/a, Warsaw 02-097, Poland.

Acta Neurochirurgica, Suppl. 50, 122–127 (1990)

The Zürich Amygdalo-Hippocampectomy Series: A Short Up-date

H. G. Wieser[1], **A. M. Siegel**[1], and **G. M. Yaşargil**[2]

[1] Neurological and [2] Neurosurgical Departments, University Hospital Zürich, Switzerland

Summary

Selective amygdalohippocampectomy was developed as a surgical treatment for temporal lobe epilesy with a well-defined unilateral mesiobasal limbic seizure onset. Since 1975, 236 patients have been operated on in Zürich. We briefly summarize recent studies on the seizure outcome with analysis of the postoperative long-term fluctuations in relation to postoperative anticonvulsant drug treatment, on the underlying neuropathology, and on the relationships between magnetic resonance scanning estimates of total volume of the removed tissue and the resection scores of amygdala, hippocampus and parahippocampus gyrus.

Keywords: Temporal lobe epilepsy; amygdalohippocampectomy; morphological findings; follow-up; results.

Introduction

On several occasions we have reported the preconditions, rationale[5–8], operative technique[8, 9], and current results of this operation. The present up-dated report considers a total of 236 operated patients (another eight are waiting for this operation).

Histological data have been analyzed for 204 patients, and follow-up data with respect to postoperative seizure outcome have been recently re-evaluated for 177 patients with a minimum follow-up of 1 year[3]. Table 1 summarizes the most important clinical data of this series.

Results

Neuropathology

There were 24 benign tumours (WHO Group I: pilocytic astrocytoma 10; ganglioglioma 8; subependymal giant cell astrocytoma (tuberous sclerosis) 4; meningioma (one fibroblastic, one meningotheliomatous) 2). Forty-six patients had semi-benign tumours, classified as WHO Group II (fibrillary astrocytoma 23; oligodendroglioma 15; oligo-astrocytoma 6; gemistocytic astrocytoma 1, ganglioglioma 1). Thirty-six had malignant tumours (anaplastic astrocytoma WHO III-19; glioblastoma multiforme WHO IV-12; anaplastic meningioma WHO III-2; anaplastic oligo-astrocytoma

Table 1. *Synopsis of the Most Important Clinical Data of the Zürich Selective Amygdalohippocampectomy Series*

	Total	Lesional	Non-lesional*
Number of patients	204	137	67
	(100%)	(67%)	(33%)
Female : male	90 : 114	60 : 77	30 : 37
	(44%) : (56%)	(44%) : (56%)	(45%) : (55%)
Right : left	109 : 95	70 : 67	39 : 28
	(53%) : (47%)	(51%) : (49%)	(58%) : (42%)
Age at seizure onset (years)	18.9 ± 15.5	23.6 ± 16.4	9.4 ± 7.0
Years with recurrent seizures prior to surgery	10.7 ± 10.2	6.3 ± 7.4	19.9 ± 9.1
Age at operation (years)	29.7 ± 13.9	29.9 ± 15.7	29.3 ± 9.3

Mean ± S.D.

* Included are the two patients with assumed Rasmussen's encephalitis.

Table 2. *Survival Time of Patients with Tumours (in months) (including 2 patients with CJD)*

Histology (WHO)	Number of patients	Post-op. survival time			
		Mean	SD	Min	Max
Anaplastic astrocytoma (III)	16	20.7	16.3	3	52
Glioblastoma multiforme (IV)	11	10.3	7.5	3	27
Fibrillary astrocytoma (II)	4	43.3	19.9	14	57
Ganglioglioma (II)	1	39			
Anaplastic meningioma (III)	1	10			
Pilocytic astrocytoma (I?)	1	39			
Anapl. oligo-astrocytoma (III)	1	15			
Gemistocytic astrocytoma (II)	1	27			
Primitive neuroectodermal tumour	1	5			
Creutzfeldt-Jakob disease (CJD)	2	26.5	19	13	40
Total	39	20.9	16.7	3	57

Table 3. *Seizure Outcome Following Selective Amygdalo-hippocampectomy at Different Postoperative Times*

Follow-up		Outcome (number of patients)				%			
		I	II	III	IV	I	II	III	IV
1 year	(n 177)	115	9	20	33	65%	5%	11%	19%
2 years	(n 138)	82	9	17	30	59%	7%	12%	22%
3 years	(n 106)	63	4	13	26	59%	4%	12%	25%
4 years	(n 91)	57	4	12	18	63%	4%	13%	20%
5 years	(n 73)	43	5	9	16	59%	7%	12%	22%
6–8 years	(n 34)	20	1	4	9	59%	3%	12%	26%
9–13 years	(n 10)	7	1	1	1	70%	10%	10%	10%

WHO III-1; anaplastic ganglioglioma WHO III-1; primitive neuroectodermal tumour 1). There were 21 arteriovenous malformations; 4 cavernous angioma, 5 hamartoma, and 1 epidermoid (total: 31). In two patients a "Rasmussen's encephalitis" was assumed.

Four patients had severe, 6 moderate and 21 slight hippocampal gliosis (total 31; many of them had "hippocampal herniation"). No histological lesion was detected in 34 patients (in 3 of them, however, intraoperatively a "hippocampal herniation" was assumed). In the so-called "non-lesional group" (n = 67) amygdalohippocampectomy was preoperatively classified as "causal" for 37 patients and for 30 patients as "palliative", e.g. amygdala and hippocampus were resected because there was proof, from presurgical evaluation with intracranial electrodes, that these structures served as important secondary pacemakers.

Follow-up

Twenty-two patients had one (20 patients), two (1 patient) or three (1 patient) re-operations because of recurrence of tumour.

Thirty-nine patients in this series have died with recurrences of their tumours (37 patients) or after having been innoculated with Creutzfeldt-Jakob disease (2 patients). The circumstances of this latter extraordinary complication have been summarized elsewhere[1]. The mean postoperative survival time of these 39 patients was 20,9 months (see Table 2).

Seizure Outcome

Table 3 summarizes the epileptological outcome data in relation to the post-operative follow-up period.

Table 4. *Antiepileptic Drugs Taken 1 Year After Operation*

Anticonvulsant drug regimen	Number of patients (%)		Outcome			
			I	II	III	IV
No drug	28	(16%)	28	0	0	0
1 drug	82	(46%)	57	6	8	11
2 drugs	43	(24%)	22	2	7	12
3 and more drugs	24	(14%)	8	1	5	10
Total	177	(100%)	115	9	20	33

Table 5. *AEDs Given in Monotherapy and Combination at Time 1 Year After Operation*

Type of drug	Monotherapy number of patients (%)		Combination of 2 AEDs number of patients (%)	
Phenytoin	46	(56%)	34	(79%)
Carbamazepine	17	(21%)	16	(37%)
Phenobarbital	15	(18%)	15	(35%)
Primidone	3	(4%)	4	(9%)
Valproate	–		11	(26%)
Diazepam and others	1	(1%)	6	(14%)
	82	(100%)	43	(100%)

Table 6. *Comparison of Results Obtained with "Causal" and with "Palliative" Amygdalohippocampectomy*

	Outcome				Follow-up (mean, months)
	I	II	III	IV	
Causal AHE (n 30)	22 (73%)	6 (20%)	–	2 (7%)	38
Palliative AHE (n 28)	3 (11%)	–	9 (32%)	16 (57%)	53

Drug Therapy

Preoperatively all operated patients were drug-resistant and had as a rule high doses of antiepileptic drugs (AEDs) in various combinations. The AED regimen of the patients at 1 year after operation is given in Table 4.

Over the whole follow-up period in 49 patients (out of 177; = 28%) the AEDs could be withdrawn after a mean postoperative treatment period of 20 months. In these patients the mean further follow-up period without AEDs is 38,8 months: 46 patients remained seizure-free (AED withdrawal at 20,1 months; follow-up without AEDs 40,2 months; means). Two patients had a recurrence of seizures following withdrawal of

AEDs: in one patient seizures persisted despite immediate de novo AED treatment (AED withdrawal 6 months after surgery, postoperative follow-up 1 year). The other patient became seizure-free following de novo AED treatment (AED withdrawal 6 months postoperatively; postoperative follow-up $5^1/2$ years). In one patient, following AED withdrawal $3^1/2$ years postoperatively, rare auras with a frequency of 1–2/year recurred; this patient preferred to remain without AEDs (follow-up without AEDs 2 years).

Table 5 provides the details of the type of AED treatment (monotherapy and combination of two AEDs) 1 year after operation.

The most frequent combinations of 2 AEDs were: Phenytoin and phenobarbital was combined in 23%,

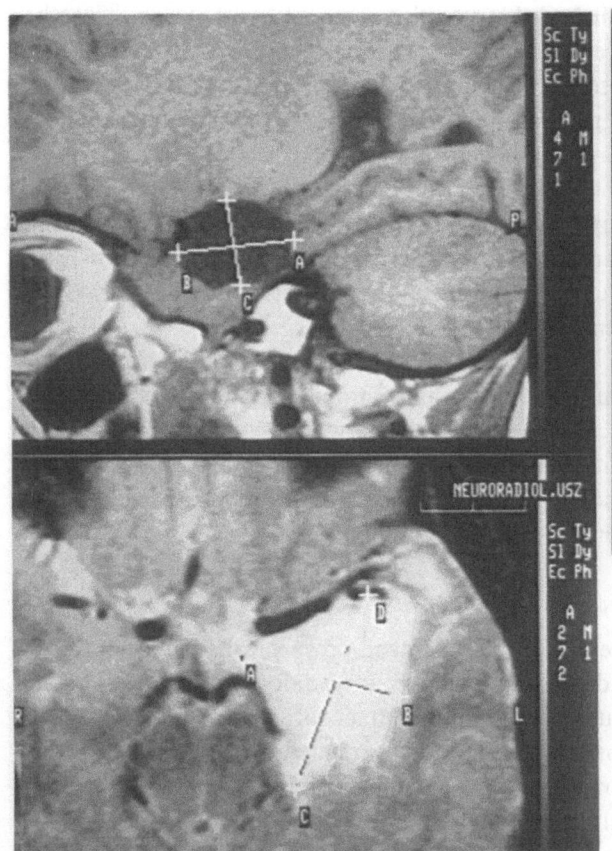

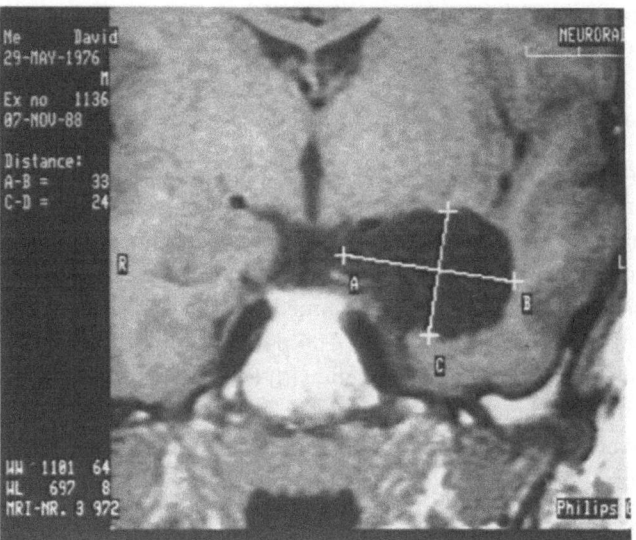

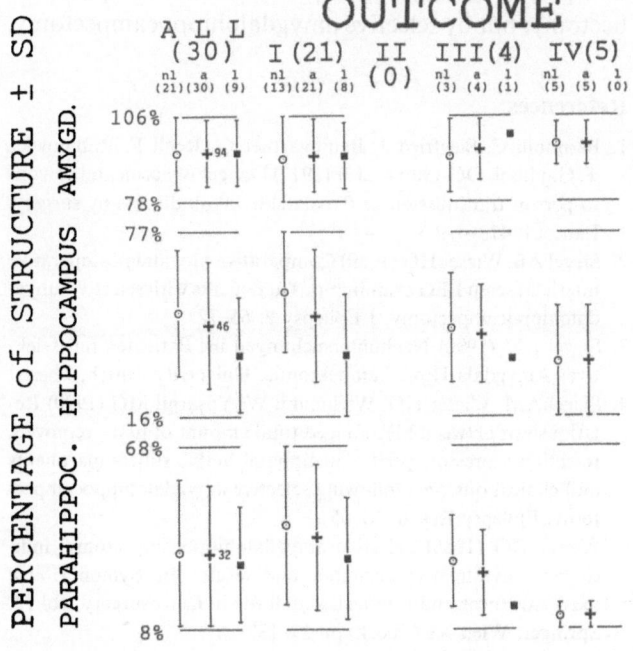

Fig. 1. MR-images (most representative sections in all three planes) showing the size and the topography of the resection following a selective amygdalohippocampectomy in an 11-year-old boy with a small oligodendroglioma (WHO II). Outcome: seizure-free since operation; postoperative follow-up 29 months; withdrawal of AEDs 11 months after surgery

Fig. 2. Mean resection scores of the most important limbic subcompartments amygdala, hippocampus and parahippocampal gyrus (in percents of the anatomical structure under consideration – for anatomical definitions see ref.[4]) and their relation with the postoperative seizure outcome I–IV. Seizure outcome is classified as follows: I seizure-free, II rare seizures (not more than 2/year), III worthwhile improvement (⩾ 90% seizure reduction *and* marked improvement of quality of life), IV no worthwhile improvement.
Numbers in brackets refer to number of patients studied. Abbreviations: *nl* non-lesional; *a* all; *l* lesional. Means and standard deviation. Note that there is a trend in favour of a better outcome in those patients with a more radical removal of the anterior parahippocampal gyrus

phenytoin and carbamazepine in 19% and phenytoin and valproate in 19%.

"Non-lesional" Group

The present up-dated report includes a total of 67 "non-lesional" patients, in whom no tumours, dysplasia, malformations and no vascular pathologies were detected (included are, however, patients with hippocampal gliosis and 2 patients with assumed Rasmussen's encephalitis). Within the total of 204 patients 44 patients had been studied preoperatively with Stereo-EEG and 43 with foramen ovale electrodes.

Overall the early follow-up results did not change substantially, with time: the outcome of this type of surgical therapy yields satisfactory *long-term* results. Postoperative seizure outcome was better in those patients where the initial seizure-onset locus had been proved to be within the resected structures. As might be expected, the results of "palliative" amygdalohippocampectomy (n = 30) are less good than in the "causally" operated group (n = 37). Table 6 compares the results obtained within these two sub-groups.

Nevertheless palliative amygdalohippocampectomy is an option in those cases in whom the primary focus lies in or close to indispensible cortex and in whom the secondary pacemakers role of the amygdala-hippocampus complex has been proven. Etiology, the presence of structural abnormalities (in particular hippocampal sclerosis), age at seizure onset, pre-operative duration of the seizure illness, and the presence or absence of postoperative EEG "spikes"[2] are further factors which influence the outcome data. More comprehensive neuropsychological data are now available[3]. In general they confirm our earlier observations that, postoperatively, seizure-free patients improve in their performance. In particular, learning and memory performance increases for material specific for the non-operated hemisphere.

Volume of Resected Tissue and Outcome

In a recent study, 30 patients in this series had special pre- and postoperative MR imaging and postoperative follow-up after at least 1 year. The total size of the resection, and the resection scores of specific limbic subcompartments were measured. These measures were correlated with the postoperative outcome. The mean volume of the removed tissue in selective amygdalohippocampectomy was 7.2 cm^3. The amygdala was removed almost completely, whereas only the anterior half of the hippocampus (mean 46%) and about one third of the parahippocampal gyrus (mean 32%) has been removed in this subpopulation. Although a small resection did not exclude a good operative outcome, in general the tendency emerged that a better outcome was obtained from a larger resection. In particular a positive correlation emerged between good postoperative outcome and the radicality of the removal of the parahippocampal gyrus, supporting our previously formulated "amplifier hypothesis" of the parahippocampal gyrus[4,5].

Figure 1 illustrates the resection in a "lesional" amygdalohippocampectomized patient, and Fig. 2 depicts the correlation between resection scores of the limbic subcompartments (amygdala, hippocampus, and parahippocampal gyrus) and postoperative outcome.

Further Developments

Recent developments in presurgical evaluation of candidates for selective amygdalohippocampectomy include (1) the multicontact foramen ovale electrode recording technique from mediobasal aspects of both temporal lobes, (2) the "selective" temporal lobe Amytal procedures, (3) more comprehensive pre- and postoperative SPECT (HMPAO; Iomazenil) and ^{18}FDG-PET examinations. The latter functional imaging methods, however, do not substitute for detailed neurophysiological examinations. Nevertheless they provide important additional information.

From our data we conclude that temporal lobe epilepsy with mediobasal limbic seizures should no longer be operated on by "standard" anterior temporal lobectomy, but by selective amygdalohippocampectomy.

References

1. Bernoulli C, Siegfried J, Baumgartner G, Regli F, Rabinowicz T, Gajdusek DC, Gibbs CJ Jr (1977) Danger of accidental person-to-person transmission of Creutzfeldt-Jakob disease by surgery. Lancet i: 478–479
2. Siegel AB, Wieser HG (1989) Comparative pre- and postoperative interictal scalp EEG examinations in patients with selective amygdalohippocampectomy. J Epilepsy 2: 65–72
3. Siegel AM (1990) Nachuntersuchungen bei Patienten mit selektiver Amygdala-Hippokampektomie. University Zürich, Thesis.
4. Siegel AM, Wieser HG, Wichmann W, Yaşargil MG (1990) Relationships between MR-imaged total amount of tissue removed, resection scores of specific mediobasal limbic subcompartments and clinical outcome following selective amygdalohippocampectomy. Epilepsy Res 6: 56–65
5. Wieser HG (1986) Selective amygdalohippocampectomy: indications, investigative technique and results. In: Symon L et al (eds) Advances and technical standards in neurosurgery, vol 13. Springer, Wien New York, pp 39–133

6. Wieser HG, Yaşargil MG (1984) Selective amygdalohippocampectomy as a surgical treatment of mesiobasal limbic epilepsy. Surg Neurol 17: 445–457

7. Yaşargil MG, Wieser HG (1987) Selective amygdalohippocampectomy at the University Hospital, Zürich. In: Engel J Jr (ed) Surgical treatment of the epilepsies. Raven Press, New York, pp 653–658

8. Yaşargil MG, Wieser HG (1987) Selective microsurgical resec-

tions. In: Wieser HG, Elger CE (eds) Presurgical evaluation of epileptics. Springer, Berlin Heidelberg New York, pp 352–360

9. Yaşargil MG, Teddy PJ, Roth P (1985) Selective amygdalohippocampectomy: operative anatomy and surgical technique. In: Symon L *et al* (eds) Advances and technical standards in neurosurgery, vol 12. Springer, Wien New York, pp 93–123

Correspondence: Prof. Dr. med. H. G. Wieser, Neurologie/EEG, Universitätsspital, CH-8091 Zürich, Switzerland.

Acta Neurochirurgica, Suppl. 50, 128–130 (1990)

Cortical Resections Outside the Temporal Lobe for Intractable Epilepsy – Excluding Multilobar Resections and Hemispherectomy

C. E. Polkey

Neurosurgical Unit, Maudsley Hospital, London, U.K.

Summary

The principles underlying such surgery are reviewed, together with details of the necessary investigation, and a review of long term results from the literature. Cortical Resection for epilepsy remains a useful treatment for drug resistant epilepsy when a cortical focus can be identified by neurophysiological or radiological means. Such a resection has a reasonable chance of relieving or ameliorating the epilepsy with only a small chance of producing or worsening any neurological deficit.

Keywords: Epilepsy; drug resistance; cortial resection; results.

Although temporal lobe resections and major resections form the majority of operations for drug resistant epilepsy, comprising 67% in the Montreal material[9] and 80% in that from the Maudsley Hospital, nevertheless a substantial number of cortical resections are performed outside of these areas. The commonest area for operation in this group is the frontal lobe (11–18%) and the least common, possibly because it is in some way resistant to epileptogenesis, is the occipital lobe. The low frequency of operations in the centro-parietal areas is perhaps determined by the fact that they are eloquent areas in which the possibility and effect of neurological deficit is greater than elsewhere.

Nevertheless the criteria for operation in these areas are the same as those applied to drug-resistant epilepsy in the temporal lobes. These, which have been set out by Rasmussen and others[1], embrace two principles, namely that the patient should have had a thorough trial of anticonvulsant medication and that a focal origin for the epilepsy can be demonstrated in an area of brain that can be removed without producing significant neurological deficit and by implication a significant increase in preexisting neurological deficit.

The patients are assessed on the basis of their clinical features including the ictal pattern, which should be consistent, and the inter-ictal history which may suggest the nature of the underlying pathology. The clinical features of frontal lobe seizures are bizarre and may need considerable experience to recognise them. A recent review by Williamson[12] includes very helpful descriptions of the various types of frontal lobe seizure. Depth electrode studies have shown that adversive movements of the head and eyes are a poor lateralising feature[10].

In *frontal lobe cases* the neurophysiological investigations may not be straight-forward. When only scalp electrodes are used little of localising value may be recorded, even with ictal telemetering, because the medial surface of the frontal lobe is inaccessible to them. On the other hand, gross unilateral frontal lobe lesions may be accompanied by bilateral EEG changes. In some cases chronic recording with depth electrodes may be necessary. This method was pioneered originally by Bancaud and his group in Paris[11] but is now commonly used. A good summary of the reasons for its use and possible problems is given by Ojemann and Engel[5].

Modern brain imaging methods using high quality CT scanning or MRI imaging have improved the demonstration of small discrete lesions within the frontal lobes.

Frontal lobe resections will vary in size between small local cortical resections and large frontal lobectomies, and in their location depending upon the site and nature of the pathology or the neurophysiological findings. When there is a clear abnormality the size and position of the resection can be guided by that abnormality with corticographic control. The resection may be limited by corticographic findings, where the pathology is obscure or not easily demonstrated upon the exposed cortical surface, or by the practical size of a frontal lobe resection which should leave one gyrus anterior to the pre-central gyrus, and respect Broca's area in the dominant hemisphere. The technique of resection is that described from Montreal in which gentle subpial dissection down to the white matter is carried out leaving the unresected banks of cortex protected by the pia-arachnoid. Some neurosurgeons find an ultrasonic dissector, such as the Cavitron useful for this purpose.

The efficacy of frontal lobe resections varies, but as with temporal lobectomy, the epilepsy is more likely to be relieved if the resected specimen contains an identifiable pathological lesion, provided it is solitary. In some patients, where the pathology is of a heterotopic or dysplastic nature, it may appear to be solitary but the epilepsy persists after the resection and one may have to assume that there are other areas of abnormality which have not been disclosed by brain imaging techniques. The range of pathology encountered in frontal lobe resections is different from that seen in temporal lobe resections. Rasmussen[7] reporting 345 frontal resections over 45 years notes that 97 (28%) had tumours, 5 (1.4%) AVM's and 167 (48%) scarring, often caused by head trauma.

The results of frontal resection are satisfactory. The largest series from the Montreal Neurological Institute report 212 patients followed for 2 to 39 years[7]. There were 22 patients (10%) who had been absolutely seizure free and a total of 76 patients had suffered less than one seizure per year of follow-up, (35%) with a further 42 patients experiencing a significant reduction of seizures so that a total of 118 patients (55%) had benefited from the surgery. There has been no mortality in the last 160 operations and no morbidity. There are minor intellectual sequelae to frontal lobe resections.

The number of patients submitted to *surgery in the centroparietal regions* is low because this is an eloquent area and in the occipital area because focal epilepsy rarely affects this area. The focal motor and focal sensory seizures which emanate from the central and parietal areas are easy to identify. Occipital seizures may begin with visual auras.

As with frontal lobe epilepsy, sophisticated brain imaging techniques have also aided the identification of lesions in the centro-parietal and occipital regions.

The neurophysiological investigation of these patients has become more sophisticated. Scalp recordings may be of limited value because they fail to resolve the dilemma in the central area of which cortical areas subserve important motor and sensory functions. Previously such information was obtained at operation under local anaesthetic, using stimulation and recording techniques which have been well described[6]. However this method was uncomfortable for both the patient and the surgeon and could not be applied to children. The neurophysiological information was limited in time and the stimulation results were brief and difficult to obtain. A number of authors notably Goldring and Gregorie[2] have described the use of mat electrodes for chronic recording and stimulation over some days after insertion under general anaesthetic and these methods can be applied to children. By these techniques information about the origin of seizures in these sensitive areas and motor and sensory mapping can be carried out in a leisurely and detailed fashion. Goldring[3] reported cortical resections from the central areas after assessment of the seizure origin and identification of motor and sensory cortex using epidural mat electrodes.

The dilemma with resections in the centro-parietal areas is the balance between producing neurological deficit and relieving the epilepsy. Rasmussen[8] in discussing the surgical technique notes that whereas the face area of the pre-central 1 or post-central cortex can be resected with little deficit, and the leg area of the post-central cortex with little practical disturbance to the patient, removal of the pre-central leg area is seldom advisable and if undertaken may result in a profound flaccid paralysis of the limb. Likewise the pre-central and post-central arm-hand areas should not be resected if there is no pre-operative neurological deficit. Resection in these areas has to be tailored, often by operation under local anaesthesia, to avoid worsening a preexisting deficit or introducing one. In occipital operations there may be a high risk of producing a homonymous hemianopia.

An elegant solution to the dilemma presented by resection of the precentral and postcentral gyri has been described by Morrell[4]. This is the operation of multiple subpial transection which is based on the idea that epileptic discharges propagate horizontally whereas the impulses controlling voluntary movement propagate vertically. If a series of vertical cuts are made in the motor cortex the epilepsy will be controlled and normal function unaffected. In 32 patients, there has been no practical neurological deficit although subtle changes could be detected on formal examination.

Rasmussen[8] reports 68 patients undergoing central resections in 40 years. Eighteen percent were rendered absolutely fit-free, 45% were down to less than one fit per year of follow-up, and overall 59% were significantly improved. In parietal resections, of which there were 132 cases the results were almost identical. Goldring[3] reports the results of extra-temporal resections in 40 patients; 18 (45%) became fit-free and a further 8 obtained a significant improvement giving an overall success rate of 65%. Morrell *et al.*[4] report that, in 20 patients undergoing multiple subpial transection and then followed for more than five years, there has been complete seizure control in 55%. Seizures recurred in some patients with progressive pathology such as Rasmussen's disease but always away from the treated area. In a number of the 40 cases reported by Goldring unsuspected indolent gliomas were found. In Rasmussen's material 266 patients underwent central or parietal resections, 103 patients had tumours (38.7%) and in 110 (41.3%) some other pathology leaving 45 (20%) with no specific pathology.

Rasmussen[8] describes 23 patients undergoing occipital resection and notes that 68% achieved a singificant reduction in their epilepsy.

References

1. Andermann F (1987) Identification of candidates for the surgical treatment of epilepsy. In: Engel J (ed) Surgical treatment of the epilepsies. Raven Press, New York, pp 51–70
2. Goldring S, Gregorie EM (1984) Surgical management of epilepsy using epidural mats to localise the seizure focus. Review of 100 cases. J Neurosurg 60: 457–466
3. Goldring S (1987) Surgical management of epilepsy in children. In: Engel J (ed) Surgical treatment of the epilepsies. Raven Press, New York, pp 445–464
4. Morrell F, Whisler WW, Bleck TP (1989) Multiple subpial transection: A new approach to the surgical treatment of focal epilepsy. J Neurosurg 70: 231–239
5. Ojemann GA, Engel J (1987) Acute and chronic intracranial recording and stimulation. In: Engel J (ed) Surgical treatment of the epilepsies. Raven Press, New York, pp 263–288
6. Penfield WD, Jasper H (1954) Epilepsy and the functional anatomy of the human brain. Little Brown, Mass
7. Rasmussen T (1975 a) Surgery of frontal lobe epilepsy. In: Purpura DP, Penry JK, Walter RD (eds) Advances in neurology, Vol 8. Neurosurgical management of the epilepsies. Raven Press, New York, pp 197–205
8. Rasmussen T (1975 b) Surgery for epilepsy arising in regions other than the temporal or frontal lobes. In: Purpura DP, Penry JK, Walter RD (eds) Advances in neurology, Vol 8. Neurosurgical management of the epilepsies. Raven Press, New York, pp 207–226
9. Rasmussen T (1987) Cortical resection for multilobe epileptogenic lesions. In: Wieser HG, Elger CE (eds) Presurgical evaluation of epileptics. Springer, Berlin Heidelberg New York, pp 344–351
10. Robillard A, Saint Hilaire JM, Mercier M, Bouvier G (1983) The localising and lateralising value of adversion in epileptic seizures. Neurology 33: 1421–1422
11. Talairach J, Bancaud J, Szilka G, Bonis A, Geier S, Vedrenne C (1974) Approche nouvelle de la neurochirurgie de l'epilepsie. Methodologie stereotaxique et resultats théraupeutiques. Neuro-chirurgie, Suppl 2
12. Williamson PD, Spencer DD, Spencer SS, Novelly RA, Mattson RH (1985) Complex partial seizures of frontal origin. Ann Neurol 18: 497–504

Correspondence: C. E. Polkey, M.D., Neurosurgical Unit, Maudsley Hospital, De Crespigny Park, London SE5 8AZ, U.K.

Acta Neurochirurgica, Suppl. 50, 131–133 (1990)

The Place of Hemispherectomy and Major Cortical Resection in the Control of Drug Resistant Epilepsy

C. E. Polkey

Neurosurgial Unit, Maudsley Hospital, London, U.K.

Summary

The results of hemispherectomy are very satisfactory but depend upon the technique used. Recent reports from a number of centres confirm that 73% of patients are fit free following hemispherectomy. Delayed pressure complications of hemispherectomy are discussed, together with the modified techniques required to cope with such problems.

Keywords: Epilepsy; drug resistance; hemispherectomy; results; late complications.

In some patients the disease process causing the epilepsy is so gross or widespread as to justify large cortical resections involving two or more lobes of the brain or even hemispherectomy. In such patients where the central areas are involved, or where hemispherectomy is contemplated, there is invariably a preexisting major neurological deficit. The need to perform multilobar resections other than hemispherectomy is relatively uncommon. In the Montreal series, 3.6% of patients underwent such resections and a further 7.4% some form of hemispherectomy. Rasmussen also reports that in this group of major resections, where about half the hemisphere remained, a lower number of patients, about 40%, remained fit free after operation when compared with subtotal or total hemispherectomy. He summarises the position by noting that although there were thought to be good reasons at the time for the lesser removal it is clearly less efficient and that, where possible, the removal of damaged areas should be as complete as possible[10].

The history and variations of hemispherectomy, or more properly hemidecortication, are of great interest. First proposed as a cure for malignant cerebral tumours by Dandy and L'Hermitte in 1928 the procedure fell into disuse for obvious reasons and the first reported use of the operation for infantile hemiplegia and seizures was from Canada[7]. However the first substantial series of hemispherectomy for HHE syndrome, namely hemiplegia, hemi-atrophy and epilepsy was by Krynauw in 1950[6]. At that time, and for some years afterwards the majority of patients subjected to hemispherectomy were victims of sudden and substantial insults to one hemisphere in the form of perinatal damage or severe neonatal infective illness leading to either major venous or arterial insufficiency followed by infarction of a large area of cerebral cortex. However it is interesting to note that the indications for operation in Krynauw's series included not only the intractable epilepsy but also the onset of intellectual deterioration. He had noted that in these patients serial EEG recordings would show that the EEG in the unaffected hemisphere would become progressively more abnormal at the same time that their intellectual performance and behaviour also deteriorated[6]. In these details lie not only the therapeutic good of hemispherectomy, but also the intrinsic interest of the disease processes which lead to the final situation in which hemispherectomy is an acceptable solution. Not only is there information about the reversibility of abnormal neurophysiological changes but also information about the way in which the maturing brain adapts its intellectual performance to major neuronal loss. The results of these major insults depend upon the patient's age and the rapidity of their onset. Even the physical disability can vary. In one patient of mine where the vascular insult was sustained in utero, the patient had voluntary finger flexion and extension before hemispherectomy, and these movements were retained afterwards. There is a complex literature, spanning many years dealing with the intellectual consequences of these events. It fails to resolve the question of whether language is initially bilateral and then gradually lateralises to one hemisphere, so that early destruction of the language area in either hemisphere leads to a temporary disability which rapidly recovers; or whether in most individuals it is always lateralised to the left hemisphere but will transfer to the right hemisphere if there is early damage to the left hemisphere. In practical terms however the later the insult to the dominant hemisphere the more language loss there is as was noted by Rasmussen and Milner in their elegant study of patients sustaining left hemisphere damage early in life[11].

Before discussing the clinical details of patients who are suitable for hemispherectomy it is necessary to describe briefly the kind of pathological processes involved since over the years they have affected both the application and the outcome of the operation. The patients considered for hemispherectomy fall into three groups. In the first group are those for whom the operation was originally described and who formed the majority of the patients described in the early series. These are patients who in some way suffer a sudden and massive insult to one hemisphere in utero, infancy or early childhood and who after recovery from this acute episode are found to be profoundly hemiplegic and who subsequently develop drug resistant seizures. They still form the majority of candidates for hemispherectomy. In the second group are patients with some congenital condition, which may or may not be progressive, such as Sturge-Weber syndrome, where a case can be made in patients with hemiplegia for early hemispherectomy[5]; or various ill-defined forms of megancephaly in which one hemisphere, or sometimes one to a great degree and the other to a much lesser degree, are the seat of a complex migrational disorder[13]. In the third group are patients who are apparently normal till childhood when they become the victim of an ill-understood disorder in which unilateral focal seizures, usually focal motor seizures appear, are relatively drug resistant and are followed by the onset of a hemiparesis. These patients when the condition progresses, which is not always the case, often show epilepsia partialis continuans. Such patients correspond roughly to those described by Rasmussen as having encephalitis with epilepsy[9]. Because there are currently no diagnostic or radiological markers which are specific for this condition, in the absence of a pathological specimen, it has to be a provisional diagnosis.

Clinically the patients who are suitable for hemispherectomy are those in whom there is an effective hemiplegia, evidence of unilateral hemisphere disease, and uncontrollable epilepsy, by whatever means this state is reached. The term effective hemiplegia requires some explanation. In patients with a sudden insult the hemiplegia is invariably present from the acute event and the seizures may appear later. However in the other two groups the hemiplegia may develop with the seizures and even after their onset. In these circumstances there can be some question of whether the hemiplegia represents permanent structural change or the functional result of epilepsia partialis continuans, if present, and whether therefore the hemiplegia would resolve if the fits could be controlled. Also in patients where such events begin relatively late there is the question of whether speech will be affected by the hemispherectomy as we have already discussed. In these circumstances consideration should be given as to whether Morrell's procedure of multiple subpial transection is appropriate[8]. If this operation is unsuccessful then hemispherectomy can be performed at a later date.

Most patients for consideration of hemispherectomy have focal motor seizures but in fifty percent there may also be generalised seizures or drop attacks or partial seizures as well. The degree of physical disability which follows hemispherectomy depends upon the underlying disease process. In practical terms the changes are slight and a fair price for the benefits of the operation. In some patients there may be a hemianopia after operation, especially those in the second two groups. With rare exceptions both the upper and lower limbs on the hemiplegia side may become stiffer and lose some movements. However it is rare for the patients to be unable to walk after operation. Most of the subjects are children or adolescents and after 20 years of age rehabilitation becomes more difficult and may present a considerable problem in the mid-thirties or older. In practice the operation makes little difference to the mobility of younger patients although the pattern of locomotion may be more cumbersome.

The intellectual sequelae of hemispherectomy are equally complex. In most series about 50% of patients have unmeasurable IQ's prior to operation. If the first group is considered, there is usually no deterioration in the intellectual performance as a result of the operation; indeed if it is successful there may be an improvement in intellectual performance, as recently detailed for a small series by Beardsworth and Adams[2]. However in the second and third groups the outcome may be different since there may still be some intellectual function in the diseased hemisphere, which has taken part in the learning of the maturing brain prior to the onset of the disease process. Many years ago

Rasmussen and Milner showed that the intellectual effects of unilateral hemisphere disease are graded, even though the insult may occur in childhood[11].

The results of neurophysiological investigation in these patients varies. The patients in the first group often have a "flat" record on the affected side with increasing abnormality in the unaffected hemisphere. On the other hand in the second and third groups there may be more obvious abnormality over the affected hemisphere.

The results of neuroradiological investigation depends upon the disease process concerned. A grossly destructive process will be apparent with gross hemiatrophy and porencephaly. Likewise Sturge-Weber syndrome will be obvious, and cerebral migrational disorder as in megancephaly, although the latter is possibly more elegantly displayed by MRI scanning than by CT scanning. In Rasmussen's encephalitis the neuroradiological findings depend upon the stage of the disease and the rapidity of its progression. If the disease has progressed severely, or over a long period of time, then there will be clear atrophy but in cases examined early in the disease there may be abnormal findings, although it is said that PET scanning shows hypoperfusion in the affected hemisphere.

The technique of hemispherectomy has been subject to change over the years because of the complication known as late delayed bleeding or by the Montreal group as late pressure complications. Classical hemispherectomy applied to patients with an infantile hemiplegia and drug-resistant seizures, who often had a behaviour disorder and were dementing, was very successful with an 80% seizure relief rate and an improvement in the behaviour. This was soon confirmed by others[12]. Within 10 years or so of the first reports of the operation it became clear from numerous reports[4] that there was a serious morbidity and mortality in between one third and one quarter of patients from late complications associated with chronic or delayed bleeding. In Montreal it was noticed that such complications occurred in 35% of patients undergoing anatomically complete hemispherectomy but was virtually absent after lesser resections. They therefore modified their practice twice. Between 1968 and 1974 they carried out only subtotal hemispherectomies, sparing the least epileptogenic area, usually either the frontal or occipital pole. This reduced the late complications but also reduced the seizure relief so that whereas 80% of patients were virtually fit-free after complete hemispherectomy only 69% were so relieved by subtotal hemispherectomy. Since 1974 they have carried out a functionally complete hemispherectomy in which the central and temporal areas including the deep temporal structures are removed and the corpus callosum divided in the frontal and parietal regions. They have 80% of patients seizure free and only one late delayed complication in 14 cases followed for between two and ten years[10]. Adams[1] modified the classical hemispherectomy technique to avoid the late complications. In a recent report they note the freedom from late complications in ten patients operated upon using this technique[2]. There was total seizure relief in 70% of these patients with improvement in behaviour and in some intellectual function similar to the results reported for classical hemispherectomy[12].

References

1. Adams CBT (1983) Hemispherectomy – a modification. J Neurol Neurosurg Psychiatry 46: 617–619
2. Beardsworth ED, Adams CBT (1988) Modified hemispherectomy for epilepsy: Early results in 10 cases. Br J Neurosurg 2: 73–84
3. Engel J (1987) Outcome with respect to epileptic seizures. In: Engel J Jr (ed) Surgical treatment of the epilepsies. Raven Press, New York, pp 553–571
4. Falconer MA, Wilson PJE (1969) Complications related to delayed haemorrhage after hemispherectomy. J Neurosurg 30: 413–426
5. Hoffman HJ, Hendrick B, Dennis M, Armstrong D (1979) Hemispherectomy for Sturge-Weber syndrome. Child's Brain 5: 233–248
6. Krynauw RA (1950) Infantile hemiplegia treated by removing one cerebral hemisphere. J Neurol Neurosurg Psychiatry 13: 243–267
7. McKenzie KG (1938) The present status of a patient who had the right cerebral hemisphere removed. Proc Am Med Assoc 11: 168
8. Morrell F, Whisler WW, Bleck TP (1989) Multiple subpial transection: A new approach to the surgical treatment of focal epilepsy. J Neurosurg 70: 231–239
9. Rasmussen T (1978) Further observations on the syndrome of chronic encephalitis with epilepsy. Appl Neurophysiol 41: 1–12
10. Rasmussen T (1987) Cortical resection for multilobe epileptogenic lesions. In: Wieser HG, Elger CE (eds) Presurgical evaluation of epileptics. Springer, Berlin Heidelberg New York, pp 344–351
11. Rasmussen T, Milner (1977) The role of early left brain damage in determining the lateralisation of cerebral speech functions. Ann NY Acad Sci 299: 355–369
12. Wilson PJE (1970) Cerebral hemispherectomy for infantile hemiplegia: A report of 50 cases. Brain 93: 147–180
13. Vigevano F, Bertini E, Boldrini R, Bosman C, Claps D, Capuo M, Rocco C, Rossi GF (1989) Hemimegancephaly and intractable epilepsy: Benefits of hemispherectomy. Epilepsia 30: 833–843

Correspondence: C. E. Polkey, M.D., Neurosurgical Unit, Maudsley Hospital, De Crespigny Park, London SE5 8AZ, U.K.

Acta Neurochirurgica, Suppl. 50, 134–135 (1990)
© by Springer-Verlag 1990

Callosotomy for the Treatment of Drug Resistant Generalized Seizures

I. Papo[1], **A. Quattrini**[2], **L. Provinciali**[3], **F. Rychlicki**[1], **M. Del Pesce**[3], **A. Paggi**[2], **F. Ortenzi**[2], **M. A. Recchioni**[1], and **B. Censori**[3]

[1] Neurosurgical Division, [2] Epilepsy Centre, Regional Hospital, [3] Department of Neurology, University Medical School, Ancona, Italy

Summary

Fifteen patients have been followed for more than one year following callosotomy having presented with long standing epilepsy, no well defined focus amenable to radical excision, and severely incapacitating atonic seizures that were refractory to anticonvulsant therapy. Atonic fits have been reduced by more than 80% in thirteen patients, with two patients suffering long term sequelae (slight dysarthria in one, and dyslexia with mild visuo-spatial disturbances in another). Anticonvulsant therapy was still required post-operatively.

Keywords: Epilepsy; drug resistance; atonic seizures; callosotomy indication results, side-effects.

Our experience with callosal section for the management of refractory non-focal epilepsy as well as the data reported in the literature have been exhaustively dealt with in some recent papers[1–3]. Therefore, in this short presentation we shall confine ourselves to updating and summarising our clinical observations and raising some unsolved questions.

Material and Methods

18 patients were submitted to callosotomy. The surgical results will be evaluated in the 15 patients with more than one year follow-up.

The criteria for selecting patients suitable for callosal section, have been the same for all patients:

- Long standing epilepsy
- Lack of well defined foci amenable to radical excision
- Proved resistance to drugs in mono or polytherapy
- Predominance of severely incapacitating atonic seizures.

In the whole series, etiological factors could be identified in 14 patients: birth anoxia in 7, birth trauma in 3, encephalitis in 1, meningitis in 1, toxic coma in 1, infantile head injury in 1. In the remaining 4 patients the etiology remains unknown. 12 patients were male and 6 female. The age of the patients ranged between 14 and 41 years (average 26.5) and the duration of the illness from 6 to 28 years (average 17.3). I.Q. was 70 in 9 and 50 in 7.

All patients exhibited atonic seizures, the frequency of which ranged between 10 and 400 per month. Most patients had 3 to 10 seizures per week. In all patients, other kinds of seizures (tonic-clonic, complex partial, simple partial) were also present. Complex partial seizures were the most common. 7 cases can be classified as Lennox-Gastaut syndromes.

On EEG examination in all cases at last two foci of spike/waves in frontal or frontotemporal regions were detected. The background rhythm was slow and poorly organised in Lennox-Gastaut cases and relatively organised in the patients with the highest IQ.

CT Scan and MRI disclosed definitely pathological features in 5 cases only.

Total callosotomy in two stages was performed in the first 2 patients only. In the remainder, 60% to 90% of the corpus callosum was divided and the splenium always spared.

Follow-up ranged from 13 to 67 months (average 29.3).

Results of Callosal Section

In this series, callosotomy proved effective in controlling atonic and tonic seizures with a sudden fall. Its effects have been less evident and unpredictable on other kinds of seizures, albeit in some cases complex partial seizures have been reduced and, occasionally, even abolished. In our patients we never observed an increase of partial seizures.

In 6 patients atonic fits have been abolished, 3 additional patients had 1–2 seizures in the whole period of follow-up (36, 22 and 13 months respectively), and in 4 patients atonic seizures have been reduced by 80%. Finally, in 2 patients the result was unsatisfactory.

From the neuropsychological standpoint, after callosotomy short-lasting cognitive impairment was observed in all patients. Transient aphemia or mutism, clearing up in 2 to 4 weeks, took place in 2 patients. The only permanent disorders have been: slight dysarthria in 1 case and dyslexia and mild visuo-spatial disturbances in 1 patient in whom total callosotomy was carried out.

Drug therapy was never discontinued after operation. In none of our patients was the drug regime significantly reduced. It is worth noting that, in 3 patients, drugs which had repeatedly proved ineffective in controlling atonic seizures became effective after surgery.

No definite correlation between the extent of callosal section, as revealed by postoperative MRI, and clinical results was found. Postoperative EEG tracings did not correlate with the effects of surgery. We have observed unmodified postoperative tracings in patients in whom atonic seizures have been controlled satisfactorily and, conversely, an almost normalized tracing in a patient with a poor surgical result.

Discussion

The therapeutic results obtained in our series are broadly in line with those previously reported in the literature. Callosotomy, whether partial or total, proved effective in controlling atonic seizures in a high percentage of cases.

It appears that the division of the corpus callosum renders previously drug-resistant atonic seizures drug-sensitive. The effect of callosal section on other kinds of fits are less evident and unpredictable. Permanent neuropsychological disorders seem to be rare provided that the splenium remains intact. Moreover, aggressiveness and behavioural disorders may be significantly improved after surgery.

In predicting postoperative outcome, the presence of frontal foci, IQ greater than 70, and normal or nearly normal EEG background rhythm with secondary spike/wave bisynchronism can be taken as favourable factors. Nevertheless, the indications for callosotomy are not yet precisely outlined and remain based overall on clinical data. The mechanism whereby callosotomy manages to control generalized seizures lies in the proposed inhibitory role of the corpus callosum on synchronous discharges.

Other unsolved problems are the extent of the corpus callosum division – total or partial – and its role in patients with severe mental retardation. It is unknown whether intervention in younger patients would control seizures and prevent mental deterioration.

References

1. Papo I, Quattrini A, Provinciali L, Rychlicki F, Paggi A, Del Pesce M, Ortenzi A (1989) Callosotomy for the management of intractable non-focal epilepsy: a preliminary assessment. Acta Neurochir (Wien) 96: 46–53
2. Papo I, Quattrini A, Provinciali L, Rychlicki F, Del Pesce M, Paggi A, Ortenzi A, Censori B, Recchioni AM (1989) Remarques sur la callosotomie dans le traitement de l'épilepsie pharmaco-résistant. Neuro-Chirurgie (in press)
3. Provinciali L, Quattrini A, Papo I, Del Pesce M, Mancini S (1988) Neuropsychological changes after callosotomy in drug-resistant epilepsy: a study of the short-term evolution. Acta Neurochir (Wien) 94: 15–22

Correspondence: I. Papo, M.D., Neurosurgical Division, Ospedale Regionale Torrette, I-60100 Ancona, Italy.

Acta Neurochirurgica, Suppl. 50, 136–141 (1990)

Soviet Investigations in Epilepsy

E. I. Kandel†

Neurosurgical Clinic, Institute of Neurology, Moscow, USSR

Summary

A wide ranging review of the Soviet literature on epilepsy is provided, detailing historical aspects, pathophysiology, investigations, medical treatment, and the long term results of surgical treatment, including stereotaxic intervention. The features of Kojevnikoff epilepsy are presented.

Keywords: Epilepsy; pathophysiology; diagnostic; management; USSR.

The aim of this report is to present a short review of Soviet investigations in epilepsy, which, because of the "language barrier" and for other reasons, are sometimes not very well known to the neurosurgeons of other countries.

Historical Aspects

The study of epilepsy in Russia began at the end of the XIXth century. Two great scientists should be remembered in connection with the investigation of the problem. Ivan Pavlov, the formidable physiologist, created the theory of higher nervous activity and studied the neurophysiological basis of epileptic excitation and its spread over the brain substance[37]. Vladimir Bechterew, the prominent Russian neurologist, neuro-morphologist and psychiatrist, published many papers and books in which different forms of epilepsy and their pathogenesis and clinical picture were described in detail[8]. The Russian neurologist Kojevnikoff in 1895 described a special form of epilepsy which now bears his name[30].

The surgical treatment of epilepsy began in Russia in the 1890's. In 1898 academician Bechterew organized in St. Petersburg the first neurosurgical department in the world where Ludwig Puusepp began to operate on many neurosurgical cases, including epilepsy.

In 1908 Bechterew and Puusepp published their book "Surgery in mental diseases", where much original data on the surgery of epilepsy was presented. During World War I Puusepp operated on many cases of posttraumatic epilepsy. In 1913 Razumovski[41] used to inject alcohol into the cerebral cortex in severe cases of epilepsy. Polenov in 1928 proposed a new operation termed subcortical pyramidotomy – section of the pyramidal tract with a special leukotome. The operation was also used in epilepsy[29]. Nicolas Burdenko, the Surgeon-General during World War II and the founder of modern Soviet neurosurgery published many reports on the surgical treatment of epilepsy before and after the War.

There are many laboratories, research institutes, clinics and departments in the Soviet Union, where basic, experimental and clinical investigations on the problem are performed. Two books devoted to the neurophysiological and biochemical mechanisms of epilepsy were published recently (Biniaurishwili *et al.* "Epilepsy and functional states of the brain"[10] and Pogodaev "Epileptology and cerebral pathochemistry. The theory ot etiology, pathogenesis and treatment of epilepsy"[38]. These books contain some original data and conclusions related to the problem.

Pathophysiological Considerations

It is well-known that epilepsy, as a disease, is the result of a complicated combination of two basic factors – epileptic "readiness" of the brain in general ("epileptization of the brain") and the development of one or several epileptogenic foci in various cortical and subcortical structures that generate periodic discharges of epileptic activity. The following stages in the development of epilepsy have been suggested: "epileptic neuron, epileptogenic focus, epileptogenic system, epileptic brain"[56].

The absence of epileptic attacks in the frequently encountered brain stem subcortical lesions of vascular, inflammatory, traumatic, or other aetiology does not confirm the theory of centrencephalic epilepsy[28]. Certain authors doubt the very existence of primary generalized attacks, believing that careful observation of all such patients will detect a focal phenomenon at the beginning of the seizure[49].

The epileptic focus: There are several definitions of an epileptogenic focus, but one of the most precise definitions is that of Romodanov (1980): "... not simply a group of ganglionic cells capable of producing a convulsive focus, but a dynamic, constantly active pathological structuro-functional system".

The thickness and enlargement of the basal membranes of microvessels and also swelling of the endothelial cells of the vessel wall were disclosed in epileptogenic foci by electron microscopy[58]. It was suggested that these changes disturb the permeability of microvessels and cerebral trophics in general. Attempts have been made to determine the quantitative size of an epileptogenic focus, which may vary greatly. For example, on the basis of electrosubcorticographic data, it has been established that the minimum diameter of an epileptogenic focus in the thalamus varies from 0.5 to 3 mm, and its volume varies from 3 to 20 mm^3[16].

The epileptogenic focus in the deep-seated structures of the temporal lobe has a higher rate of metabolism than the adjacent cerebral tissue, which is an indication of excessive O_2 consumption. This was revealed by implanted polarographic electrodes[68]. At the same time it was shown that in cortical as well as deep-seated epileptogenic foci, there is a marked decrease in local CBF producing hypoxia of cerebral tissue. PO_2 in the tissue of the focus is also decreased[64, 65]. In contrast, another study disclosed that the regional blood flow, investigated by the xenon method, in the brain regions of high epileptic activity is increased or decreased with about the same frequency. The blood flow in the adjacent zones of the hemisphere was markedly reduced, and reactivity of cerebral vessels was changed[24].

Quite frequently the epileptogenic focus in the temporal lobe, as time passes, leads to the formation of secondary foci in the mesencephalic reticular formation irrespective of the primary focus[61]. This may account for approximately 30% of patients with a clinical picture of temporal lobe epilepsy who have an epileptogenic focus or foci in other parts (lobes) of the brain[23].

Diagnostic Investigations and EEG Studies

Many published papers present studies of EEG phenomena investigated by using surface and implanted electrodes in different forms of epilepsy. It was noted that the rate of stable epileptic phenomena in the interictal EEG is only about 10%[3]. The typical EEG changes (spike-wave complexes) in temporal lobe epilepsy are recorded from the mediobasal temporal areas from one or both sides. Discharges from the hippocampus through the limbic circle can spread to the gyrus cinguli and be recorded by scalp EEG[17]. The correlation analysis of EEG taken by long-term monitoring with the help of radiotelemetry has shown the very complicated pathway of propagation of the discharge with consecutive involvement of different cerebral structures[3]. The spatial localization of the local and generalized epileptic activity on EEG was investigated by computer analysis of consecutive minimization of measured and estimated potentials. There is a possibility that a computer may be able to calculate the spatial localization of the activity not only on the cerebral surface, but also to measure its depth[14].

In focal epilepsy, the focus is most frequently localized in the temporal lobe (28%) or on the convexity (22%) or the medial surface of the frontal lobe (8%): 21% of patients were found to have multifocal epilepsy[23]. Although generalized convulsive attacks (grand mal epilepsy) are not commonly observed in temporal lobe epilepsy nevertheless, when the epileptogenic focus is localized in deep-seated structures of the temporal lobe, typical primary generalized attacks may occur, complicating the diagnosis[49].

The study of epileptic activity in the cortex and deep-seated structures of the temporal lobe in 65 patients with mono- and bitemporal foci during induced sleep was performed[34]. The fusiform rhythm is mainly marked by cortical representation and is not recorded in the deep temporal structures on the side of the determinant epileptic focus. The principal foci were localized in the limbic temporal structures. In bilateral independent foci with a high level of activity, focal discharges were preserved by various methods of sleep activation. These data are important for the selection of the appopriate surgical strategy. At times, in the so-called paroxysmal phase of sleep, it is possible to use implanted electrodes to observe a sharp increase in epileptic activity in deep-seated brain structures[62].

The diagnostic value of the EEG and CT were compared in a group of 100 infants and young children with different form of epilepsy (grand mal, local fits and myoclonic convulsions). Surprisingly, the presence of cerebral pathology on EEG was noted in 92% of the cases, but CT provides such information in only 35% of the cases and gives no evidence about substantial differences between various forms of fits. The data obtained have shown that the diagnostic value of the EEG is enhanced with increase of child's age, but the value of CT does not change[13].

Pharmacological activation of the epileptic attack and the study of EEG changes induced by drugs have also been studied[48, 25].

H-reflex study of patients with frequent seizures reveals both an increase and decrease in the activity of the spinal cord motoneurons. After disappearance of seizures postoperatively, the H-reflex threshold is significantly elevated. In this connection, a disturbance of the supraspinal regulation of the alpha-and gamma-motoneurons in epilepsy is suggested[5].

Using long-term electrodes implanted into different cerebral structures (frontal and parietal cortex, amygdala, thalamic nuclei) in 98 epileptics, the influence of stimulation of these structures on H-reflex and motor phenomena was studied[6]. The threshold and amplitude of the H-reflex were shown to be elevated and reached the peak before the appearance of bilateral convulsions. The clinical characteristics of the activated epileptic phenomena depend not only on the site of electrical stimulation but also on the direction of the after-action discharge propagation.

The changes in short-latency brain stem acoustic evoked potentials in posttraumatic epileptic cases were carefully studied[1].

The changes in the ventricular system in 120 epiletics were disclosed as early as during the first year of the disease with a tendency for gradual progression[55].

CSF investigation reveals a decrease in the ionized Ca concentration, especially in patients with intensive epileptic activity on the EEG[9]. This fact may be explained by an increased uptake of Ca into neurons.

The study of urinary excretion of catecholamines in epileptics has shown that excretion of DOPA does not change, of dopamine increases 2.5 times, of noradrenaline decreases 3 times and of adrenaline – 1.6 times[29, 15]. The authors suggest that changes in the catecholamine system in epilepsy may be one of the factors in the pathogenesis of the disease.

Medical Treatment

In the Soviet literature, there are no notable achievements in the drug treatment of epilepsy, but some reports may be mentioned. The study of folic acid treatment of 137 epileptic cases has shown encouraging results, particularly with regard to improvement of the neuropsychological aspects[12].

The study of folic acid in the blood plasma by radioimmunoassay in 78 epileptic patients, identified a correlation between its content and side effects during treatment[11]. A reduction of the content of folic acid to normal was followed by a reduced incidence of epileptic fits. On the contrary, the larger the concentration of blood folic acid, the greater the risk of seizure recurrence.

The medical treatment of epilepsy in children by allopurinol, which increases the serotonin content in platelets, yielded positive effects only in 10 out of 28 children having frequent seizures[18].

The favourable effect of hemabsorption on a resistant status epilepticus and serial epileptic attacks in a limited group of patients was noted[42]. High circulating levels of immune complexes in the blood serum are considered an indication for hemabsorption.

It is well known that the problem of timing of surgery for epilepsy is important. The majority of neurosurgeons believe that drug treatment should be tried for approximately 3 years[20], but there has been a tendency to reduce this period to 1–2 years[23]. Earlier surgical intervention is indicated in children[69].

Surgical Treatment

Focal Epilepsy

Soviet neurosurgeons have accumulated extensive surgical experience in the treatment of focal epilepsy. For example, in Leningrad, the Polenov Institute of Neurosurgery published the follow up results of 761 operated cases of focal epilepsy who had been under observation for 30 years. Positive long-term results were obtained in 70% of the patients; in 31% there was almost total termination of attacks and in 39% its frequency diminished considerably. In the same Institute the analysis of the results of surgery in 328 cases of epilepsy in children, 114 of whom had multifocal epilepsy, were reported[22]. In another report the results in 104 children and adolescents operated on by the open method was presented[57]. The system of diagnostic and operative treatment ensured favourable outcomes in 88% of the children with focal epilepsy.

An analysis of some of the causes for the low efficacy of surgical treatment of epilepsy was performed[61]. Among these causes the author stressed the possibility, after stereotactic and classical operations, of the existence of a reciprocal interrelationship between epileptic foci and the activation of one of them after the removal of the other.

Temporal Lobe Epilepsy

Temporal lobe resection, which was very popular about 20–25 years ago, has now sharply decreased. The difference in the results of stereotactic treatment of temporal lobe epilepsy in 40 cases in connection with unilateral and bitemporal foci was stressed. A significant postoperative improvement was achieved in 73% of cases with monotemporal lesion and only in 44% of cases with bitemporal foci[52]. A frequent complication of the operation is homonymous hemianopsia, total or partial, in 30% of the cases[31]. A pronounced Korsakoff's syndrome develops in certain cases[4].

It was noted that late relapses of attacks are possible after resection of the temporal lobe with early good results. Recurrences within a period of 3 years developed in 30% of the operated patients, and from 3 to 9 years in another 6%[52]. This late recurrence of seizures after unilateral resection of a temporal focus was studied[53]. Incomplete resection of the focus, the existence of bilateral temporal foci, and a family history of epilepsy may be the possible causes of late recurrence. Despite the recurrence of the seizures, the results of the operation should be considered satisfactory because the protracted remission after the operation ensured adequate social adaption of the patients while resection of the focus considerably reduced the resistance of the disease to anticonvulsive drugs.

The stereotactic destruction of deep temporal nuclei – amygdala or hippocampus or both is now the dominant technique in the surgery of temporal lobe epilepsy. The chief indications for a stereotactic operation in temporal lobe epilepsy are typical temporal seizures or grand mal with a clearly defined temporal component, the presence of bitemporal or multiple unilateral foci, and the presence of pronounced psychic and emotional disorders[70].

The technique and results of stereotactic treatment of temporal lobe epilepsy have been described in many papers[67, 69, 7, 43, 44, 19, 52, 51, 62, 63, 74, 54, 26, 27]. The majority of neurosurgeons consider stereotactic destruction of the hippocampus a more effective operation than destruction of amygdala.

An important modification of the stereotactic technique for destruction of the hippocampus has to be mentioned. The idea of introducing the electrode via the long axis of the hippocampus was developed by Sramka and Nadvornik[71], who described the operation as *stereotactic longitudinal hippocampotomy.*, In the opinion of these authors, the approach that they have suggested eliminates a serious defect in the Talairach technique in which, for the total destruction of the hippocampus, it is necessary to introduce several electrodes from various points on the lateral surface of the temporal lobe. The favourable results of the operation were published with Soviet neurosurgeons[32, 33].

The anatomical variability of the temporal horn, which is important for stereotactic destruction of amygdala or hippocampus, or both, was carefully studied[62]. It is necessary to take into account the great individual variability of the temporal horn. Its length varies from 21 to 36 mm; moreover, the difference in the length of the right and left horns may be as much as 7 mm, whereas the width varies from 8 to 14 mm with the two sides differing as much as 3 mm. The value of the angle between the intercomissural line and the basic temporal line, according to the data of Talairach *et al.* obtained from anatomic investigation, fluctuates between 23.5° and 37.5°. Similar calculations on ventriculograms of 40 patients demonstrated greater variability of that angle – from 17° to 52°[60].

Kojevnikoff Epilepsy

The original form of epilepsy, described by the Russian neurologist Kojevnikoff[30], is caused by a chronic CNS lesion resulting from tick-borne encephalitis. This disorder was discovered in West Siberia and studied by the Soviet scientists Zilber, Chumakov, Grashchenkov, and others in 1937–1940. Hyperkinesias in Kojevnikoff epilepsy usually affect one half of the body. Bilateral myoclonus occur in only 10% of the cases[35]. Generalized epileptic attacks represent the second typical clinical manifestation of this disease. As a rule, such seizures, although uncommon (19% of the cases), begin with an intensification of myoclonic twitching with the subsequent spread of convulsions to other extremities, developing into a generalized attack with loss of consciousness. On the basis of clinical and electrophysiological investigations, three forms of Kojevnikoff epilepsy are defined: cortical, cortico-subcortical and subcortical[35].

Clinical experience in the stereotactic treatment of Kojevnikoff epilepsy has been published[67]. In the ten patients operated on, the disease was a result of tickborne encephalitis. Myoclonic hyperkinesias in the face and extremities were combined with generalized epileptic seizures. After stereotactic destruction of the ventrolateral nucleus of the thalamus, the hyperkinesias disappeared in half of the cases and diminished in the rest. Approximately the same results were obtained with respect to generalized epileptic attacks.

Combined Operations

Some neurosurgeons believe that combined operations with open or stereotactic destruction of several cerebral structures are more effective, although naturally such operations are more complicated and longer in duration. This is evident from an interesting study based on a vast amount of clinical material[46]. The authors analyzed the results of many years of a joint study by the neurosurgical clinics in Kiev and Warsaw. In all, 440 classical and 151 stereotactic (destruction of amygdala, campus Foreli and hippocampus) operations and 48 combined operations – initially, resection of the temporal lobe, followed by stereotactic destruction of campus Foreli – were carried out over 15 years. Good results, after the classical operations, were noted in 55% of the patients, after stereotactic procedures in 39% and after combined operations in 60% and improvement, correspondingly, was seen in 31, 42, and 25%.

An interesting operative technique in multifocal epilepsy has been proposed[21]. After conventional craniotomy and the determination of the cortical epileptogenic focus with the aid of ECoG, an electrode is inserted stereotactically in to one of the thalamic nuclei or in to one of the mediobasal temporal structures projecting to the cortical focus. The subsequent surgical procedures are determined by the results of electrophysiological investigation. Based on these results, either the cortical focus is extirpated or the subcortical structure is destroyed or both are performed consecutively.

Another form of combined treatment that may be effective for bilateral epileptogenic foci is open resection of the temporal lobe on one side and stereotactic destruction of deep-seated temporal structures on the other side[70, 51].

In order to improve the results of isolated stereotactic destruction of amygdala or hippocampus some authors destroyed both structures in a one-stage operation[73, 51, 33]. It was shown that in attacks accompanied by unilateral hyperkinesias, the simultaneous destruction of amygdala or hippocampus on the side of the focus and ventro-lateral thalamic nucleus eliminates not only seizures but athetoid hyperkinesia as well[52].

In severe cases with mental disturbances, epileptic status, and diffuse atrophy of the brain, the combined operation is indicated, for instance, bilateral amygdalotomy or amygdalohippocampotomy with unilateral destruction of fornix and the anterior commissure, or sometimes with different combinations of the de-

struction of centrum medianum, VL, gyrus cinguli and anterior, ventral, or dorsomedial thalamic nuclei[47, 52, 45]. After such operations, in the severe group of patients, in the majority of cases the results were somewhat better than after destruction of only the amygdala or hippocampus.

More than 20 different combinations of stereotactic destruction of subcortical structures (amygdala, hippocampus, fornix, gyrus cinguli, centrum medianum, ventro-lateral nucleus, campus Foreli, stria terminalis) unilaterally or bilaterally were employed in various forms of epilepsy[62], because a single small lesion does not ensure a stable therapeutic effect. The author believes that such an effect can be obtained only if relatively massive lesions of several subcortical structures, especially amygdala on both sides and hippocampus on the side homolateral to the dominant epileptogenic focus, is achieved. Additionally, it is desirable to destroy those structures that propagate the epileptic discharge.

In temporal lobe epilepsy the supplementary stereotactic destruction (after amygdala destruction) of anterior commissure and fornix was recommended[52]. A good result from unilateral hypothalamotomy after ineffective bilateral amygdalotomy was also noted.

The published data indicated that combined operations with destruction of several deep-seated brain structures do not increase the frequency of postoperative complications[47, 62].

Chronic Stimulation of Subcortical Structures

A new trend in the treatment of generalized and sometimes focal epilepsy is the employment of stimulation by chronic implanted electrodes. Such attempts were prompted by numerous experimental investigations that showed that there are cerebral structures that inhibit the discharges of epileptogenic foci or prevent their dissemination to other cerebral structures[36]. Today there are no longer any doubts that such inhibiting structures are present in the human brain. First and foremost, they include the cortex and the deep-seated nuclei of cerebellum, nucl. caudatus and quite possibly centrum medianum, intralaminar nuclei, and other subcortical structures[72, 9, 25, 62, 33].

In stereotactic operations for temporal lobe epilepsy it was shown that low frequency (4–6 Hz) stimulation of one nucl. caudatus inhibits epileptic activity in both amygdalas and interrupts a convulsive fit[62]. Even after brief stimulation, such an inhibitary effect may last for several hours. A similar effect was obtained by stimulating nucl. dentatus, for which a high frequency stimulating current is recommended. A low frequency current is advisable, since high frequency stimulation of nucl. caudatus (50 Hz) may intensify epileptic activity.

To date, only a few papers have appeared in the Soviet literature about the therapeutic action of stimulating an epileptogenic focus using implanted electrodes. The mechanism of the antiepileptic action of stimulation remains unclear, and the therapeutic value of the method still cannot be considered proven. However, stimulation of the epileptogenic focus in status epileptic may, paradoxically, prove successful[9]. Several authors have reported promising results from chronic electrostimulation of nucl. dentatus, nucl. caudatus, hippocampus, centrum medianum and other structures[60, 9].

Hemispherectomy

There are few reports in the Soviet literature about the results of hemispherectomy. The long-term follow up in 12 patients with childhood residual encephalopathy and epilepsy who underwent hemispherectomy has been reported[3]. Cessation of seizures and epileptic activity on the EEG in the majority of cases was combined with a rise of the H-reflex threshold and improvement in the performance of the motor test.

References

A full bibliography is available at the author's address.

Correspondence: E. I. Kandel, M.D., D.Sc., Neurosurgical Clinic, Institute of Neurology, Academy of Medical Sciences of the USSR, Moscow 123367, USSR.

Acta Neurochirurgica, Suppl. 50, 142–144 (1990)
© by Springer-Verlag 1990

Neurosurgical Aspects of Epilepsy
A Review of the Seminar

M. Brock

Department of Neurosurgery, Universitätsklinikum Steglitz, Berlin

In ancient times, epilepsy was called "the holy disease". It remains unclear if this was so because the afflicted individual was supposed to be harbouring divine powers, or because he or she was being punished by the gods. Probably the gods alone know the answer. It appears, however, that this is not the only secret about epilepsy that the gods want to keep hidden from man. Although a considerable amount of progress has been made in virtually every field of epilepsy research, this progress has yet to answer some of the essential questions, but has unravelled a myriad of new uncertainties.

It is easier to find a definition for the epileptogenic focus than to find the focus itself. The epileptogenic focus is defined as "an area of the brain causing epileptic discharges, the removal of which is followed by disappearance (or at least reduction) of epileptogenic activity". The situation becomes ever more intricate when one considers the anatomical substrate of such a focus, be it a scar, tumour, hamartoma, vascular malformation, or remains the "secret of the gods". Such a focus, which has the physiological characteristics of specific hyperexcitability associated with focal functional deficit, undergoes a dynamic evolution characterized – at least experimentally – by three stages: the maturation stage, the convulsive stage, and the remission stage. Removal of the focus during the convulsive stage does not significantly reduce epileptogenic acitivity.

There is one more variable in this multifactorial equation: once established, epilepsy may become independent of the original lesion and be maintained by newly developed secondary foci.

Experimental studies using the pyramidal cells of the hippocampus of the rat indicate that the cell membrane is in reality a geographical map of multiple regions with quite different topical functional properties, produced by a series of excitatory and inhibitory impulses (mediated, among others, by glutamate, acetylcholine, GABA, etc.). This harmony is disrupted by experimental epilepsy. Both the membrane permeability to potassium ions and the changes in permeability evoked by such selective agonists as quisqualate, kainate, N-methyl-D-aspartate (NMDA), and endogenous agonists, such as L-glutamate and L-aspartate, facilitate entry or displacement of chloride, calcium and magnesium ions. Such complexity not only blurs the picture for the man in the laboratory but makes it virtually incomprehensible for the clinician, who is barely able to understand the message that "the modulation of potassium is critically important for cell activity".

It is common knowledge that epilepsy is associated with focal and generalized, primary and secondary alterations in tissue, blood-flow, and metabolism. The rather old concept of selective vulnerability of the brain, originally established on the basis of data gained from experiments on intrauterine and perinatal hypoxia have been refined recently: there is also a differential regional vulnerability, which depends on whether the noxious mechanism is hypoglycemia, ischemia or epilepsy. In addition, not only ionic changes but also inhibition of protein synthesis may result from epileptic acitivity. Intraoperative monitoring of cortical amino acid release by means of continuous superfusion through semipermeable membranes indicates an increase in ethanolamine and taurine release as well as a decrease in serine and glutamine release in human epileptogenic foci.

Histopathological studies of human temporal lobes resected at surgery have demonstrated that the pyramidal and granule cells of the hippocampus are reduced in number, mainly in the rostral parts of this structure. This neuronal loss appears to be selective and delayed, and is unaccompanied by gliosis. These findings raise the question as to whether selective partial ablation, restricted only to the rostral part of the hippocampus, might be as effective as more extended procedures. This hypothesis is substantiated by the fact that the best clinical results were observed in those patients whose rostral hippocampi revealed the most pronounced disappearance of pyramidal and granular cells.

The neurosurgeon is confronted more often with cases of partial rather than generalized epilepsy. It is precisely these cases of partial epilepsy that constitute a community problem both quantitatively and because they frequently fail to respond to pharmacotherapy. It is always assumed that pharmacotherapy is useful, but we have no knowledge whatsoever about the outcome of untreated epilepsy, since this disease has always been the target of medication ever since the introduction of bromide more than one century ago. Nevertheless, 20 years after having been diagnosed as epileptics, 70% of all patients have remained free of seizures for at least 5 years. It is known that the prognosis of epilepsy depends on several factors, such as the type and frequency of the seizures, the age of the patient, the underlying cerebral disorder, and the duration of epilepsy, to mention only the most important. Increasing the dosage of any antiepileptic drug or combination of drugs will not significantly reduce the probability of recurrent seizures, but will clearly increase the probability of side-effects. This is why routine antiepileptic prophylaxis remains questionable both in posttraumatic and postoperative cases.

It is clear that the incidence of posttraumatic epilepsy is related to the severity of the injury, reaching about 50% in penetrating combat injuries. Posttraumatic epilepsy is more frequent under the age of 5. It occurs within the first 24 hours following injury in 2/3 of the cases and within the first 6 hours in 60% of the children. Posttraumatic unconsciousness increases the probability of posttraumatic seizures. In addition to traumatic head injuries, certain surgical lesions, mainly meningiomas, arteriovenous malformations, aneurysms of the middle cerebral artery and brain abscesses are also associated with a high incidence of seizures believed to depend on the degree of primary cortical damage, subarachnoid blood, and cerebral manipulation at surgery. A well-controlled prospective study has demonstrated that anticonvulsant prophylaxis does not reduce the incidence of about 36% of

postoperative seizures but does bear an increased risk of side effects.

With reference to the surgical treatment of epilepsy, certain conditions must be fulfilled to justify its indication: (1) resistance to pharmacological treatment, (2) invalidating character of the seizures, and (3) origin from an organic brain lesion.

A very careful presurgical diagnosis is, of course, a prerequisite. The surgical approach in itself depends on the cause, location and physiological characteristics of the lesion and may employ one of the following strategies: (1) removal of the causative lesion; (2) removal of the causative lesion plus the epileptogenic zone; (3) removal of the epileptogenic zone alone, if no other causative lesion is found; (4) procedures, such as commissurotomies, with the aim of preventing propagation of seizures; (5) stimulation procedures to reduce the level of epileptogenicity, and (6) (still in the experimental phase) chronic local infusion of mediators or implantation of tissues producing such mediators.

The results of surgery for epilepsy are best for hemispherectomy and poorest for topectomy. In other words, the more radical the surgery, the better the results. Despite this, precise localization of a focus is essential and should be based on multiple criteria, starting from careful seizure anamnesis and neurological examination and going as far as the chronic implantation of deep-seated electrodes and stereo-EEG explorations, which should include perisylvian and opercular structures whenever the underlying pathology so requires. The ultimate aim of such explorations, and also of psychological testing, is to find out in advance the price the patient will have to pay in order to achieve freedom from seizures. In this process of presurgical screening, modern imaging methods, such as NMR with the administration of gadolinium and comparative planimetry of the temporal lobe structures may be of considerable help. In vivo flow studies with SPECT have confirmed well-known experimental observations on deep focal blood flow and, wherever available, are being incorporated in preoperative screening. The same applies to metabolic data gained from PET. Not surprisingly, each of these procedures is being considered the best and most reliable by those employing it. For those of us who have witnessed methodological development for "so many a summer", this apostolic attitude has the distinct flavour of a "déjà vu".

The main objection to the data obtained by all these forms of imaging is the lack of proven coincidence between the morphologic focus and the true point of origin of the seizures, not least of all because an epileptic focus may be a rapidly wandering phantom already distant from its original site when the patient enters the department of neuroimaging.

Of course, the ultimate goal of all our efforts, in Neurosurgery in general and in the field of epilepsy in particular, is reintegration of the pateint into a normal social, familial, and professional life. Hence rehabilitation is a vital integrating part of any therapeutic concept, and a very rewarding one, as we have seen. Unfortunately, psychosocial analysis of the patients' situation is still neglected. Neuropsychiatry is still a new field, while the perspective of health economics is completely ignored. We must improve prevention, psychiatric assessment, reeducation, training, and job counselling of the epileptic patient. At least in Europe, politicians are woefully unaware of these problems, although the total cost of invalidity pensions for the state is thousands of times higher than the cost of even the most sophisticated equipment and methodology.

Sadly, there is a large discrepancy between the large number of suitable candidates, who would most probably benefit from adequate surgical treatment and the very small number of patients actually treated in this way. This is a challenge to us all. There is virtually no disease as treacherous and as threatening to the individual as epilepsy. It is almost always unexpected. It is always unwelcome. Let us join our efforts and proceed with our work, because our patients have their eyes turned towards us and their hopes entrusted into our hands.

Correspondence: M. Brock, M.D., Department of Neurosurgery, Free University of Berlin; Klinikum Steglitz, Hindenburgdamm 30, D-1000 Berlin 45.

J. D. Pickard, F. Cohadon,
J. Lobo Antunes (eds.)

Neuroendocrinological Aspects of Neurosurgery

Proceedings of the Third Advanced Seminar
in Neurosurgical Research,
Venice, April 30–May 1, 1987

Acta Neurochirurgica / Supplementum 47

1990. 60 figures. VII, 128 pages.
Cloth DM 158,–, öS 1106,–
Reduced price for subscribers
to "Acta Neurochirurgica":
Cloth DM 142,–, öS 995,–
ISBN 3-211-82160-0

Contents: B. J. Everitt and T. Hökfelt: Neuroendocrine Anatomy of the Hypothalamus. – J. D. Vincent and G. Simonnet: Neurohormonal Communication in the Brain. – I. Assenmacher: Central Control of Circadian and Ultradian Neuroendocrine Rhythms. – J. Lobo Antunes and K. Muraszko: The Vascular Supply of the Hypothalamus-Pituitary Axis. – M. D. Page et al.: A Clinical Update on Hypothalamic-Pituitary Control. – R. Fahlbusch et al.: Clinical Syndromes of the Hypothalamus. – G. Teasdale: Pituitary Tumours: Problems and Questions. – P. Lees: Intrasellar Pressure. – I. Lancranjan: The Medial Treatment of Prolactin and Growth Hormone-Secreting Pituitary Tumours. – R. M. Buijs: Vasopressin and Oxytocin Localization and Putative Functions in the Brain. – St. Lightman: Central Nervous System Control of Fluid Balance: Physiology and Pathology. – V. Walker: Fluid Balance Disturbances in Neurosurgical Patients: Physiological Basis and Definitions. – G. Neil-Dwyer et al.: The Stress Response in Subarachnoid Haemorrhage and Head Injury. – E. F. M. Wijdicks et al.: Hyponatremia and Volume Status in Aneurysmal Subarachnoid Haemorrhage. – R. J. Nelson: Blood Volume Measurement Following Subarachnoid Haemorrhage. – T. Dóczi et al.: Central Neuroendocrine Control of the Brain Water, Electrolyte, and Volume Homeostasis. – M. Brock: The Hypothalamus: New Ideas on an Old Structure.

G. Broggi, J. Burzaco, E. R. Hitchcock,
B. A. Meyerson, S. Tóth (eds.)

Advances in Stereotactic and Functional Neurosurgery 8

Proceedings of the 8th Meeting of the European
Society for Stereotactic and Functional
Neurosurgery, Budapest 1988

Acta Neurochirurgica / Supplementum 46

1989. 58 figures. VIII, 114 pages.
Cloth DM 140,–, öS 980,–
Reduced price for subscribers
to "Acta Neurochirurgica":
Cloth DM 126,–, öS 882,–
ISBN 3-211-82120-1

The present work comprises selected papers from a much larger group of interesting and important communications to the European Society for Stereotactic and Functional Neurosurgery. They represent modern views on a wide variety of stereotactic surgical topics from internationally acclaimed experts in this field. The neurosurgeon who has little or no acquaintance with this fruitful sub-speciality will be surprised to find very broad applications of the technique which is gradually replacing many conventional neurosurgical procedures.

Contents:

Preface – Epilepsy – Spasticity and Movement Disorder – Pain – Tumours – Vascular Diseases – Technical

Prices are subject to change without notice

Springer-Verlag Wien New York

Moelkerbastei 5, P.O. Box 367, A-1011 Wien · Heidelberger Platz 3, D-1000 Berlin 33
175 Fifth Avenue, New York, NY 10010, USA · 37-3, Hongo 3-chome, Bunkyo-ku, Tokyo 113, Japan

Rezio R. Renella

Microsurgery of the Temporo-Medial Region

1989. 50 partly coloured figures.
XII, 203 pages.
Cloth DM 158,–, öS 1110,–
ISBN 3-211-82144-9

Prices are subject to change without notice

The differentiation of the temporal lobe into a lateral neocortical and a medial allocortical region is supported by developmental, anatomical and clinical evidence. Although this view of a dual temporal lobe is generally accepted by neurosurgeons dealing with functional surgery, it still receives little attention by those approaching structural abnormalities located or extending into the medio-basal region. Consequently, the characterization of the temporo-medial area as a distinct surgical region is still lacking. The major object of this study is to analyse the medial part of the temporal lobe as a distinct surgical region and to integrate the microsurgical and physiological aspects into a concept applicable to the several types of temporo-medial lesion. The study includes five sections. The first section is devoted to the morphological aspects. The second and the third sections present a simplified clinical approach to temporo-medial lesions and analyse the ancillary investigations which are indispensable for characterizing their structural and functional features. The fourth section deals with the surgical aspects of temporo-medial lesions, and especially with the selection of the optimal approach having regard to the location of a given process, and to the extent of the functional changes. The last section is devoted to commentaries concerning the neuropathological aspects and the outcome of surgery in the temporo-medial region.

Contents:

Springer-Verlag Wien New York